AF682097

Imogen
Cunningham

Flora

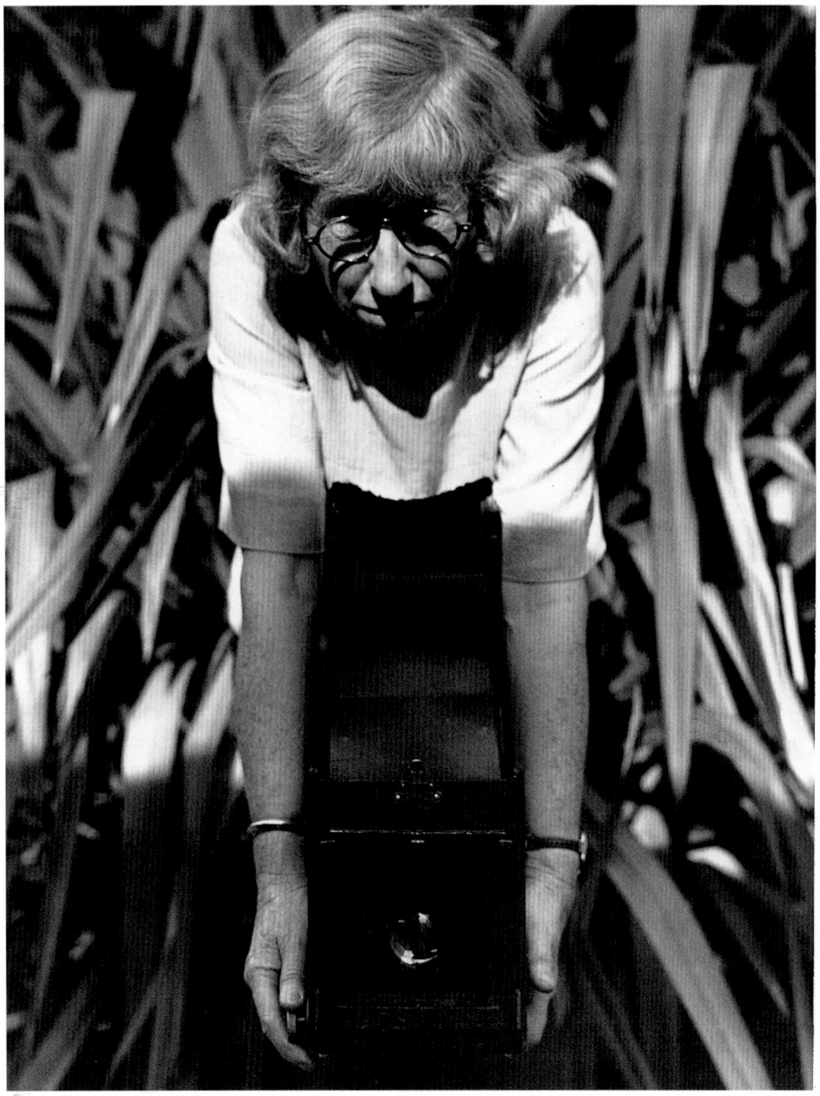

Imogen Cunningham

Flora

PHOTOGRAPHS BY IMOGEN CUNNINGHAM

TEXT BY RICHARD LORENZ

A BULFINCH PRESS BOOK
LITTLE, BROWN AND COMPANY
BOSTON • NEW YORK • LONDON

To my mother, Mary Hewko Lorenz, and the memory
of my father, Ignatius Michael Lorenz

Copyright © 1996 by Richard Lorenz

Photographs and writings by Imogen Cunningham © 1970, 1974, 1976, 1977, 1978, 1992, 1993, 1994, 1996 by the Imogen Cunningham Trust. All rights reserved. Permission to reproduce any Cunningham photograph must be obtained from the Imogen Cunningham Trust, 12 Hillcrest Court, Berkeley, California 94705.

All rights reserved. No part of this book may be reproduced in any form or by any electronic or mechanical means, including information storage and retrieval systems, without permission in writing from the publisher, except by a reviewer who may quote brief passages in a review.

First paperback edition, 2001

ISBN 0-8212-2731-9
Library of Congress Control Number 00-108540

Edited by Suzanne Kotz
Designed by Ed Marquand with assistance from Susan E. Kelly
Produced by Marquand Books, Inc., Seattle
Type set in Centaur with Arabesque ornaments
Printed in Singapore by CS Graphics

All photographs by Imogen Cunningham are in the collection of and reproduced courtesy of the Imogen Cunningham Trust. Unless otherwise noted, all illustrations reproduced herein are gelatin silver prints.

Cover: Imogen Cunningham, *Calla Leaves,* about 1930 (plate 54).
Frontispiece: Henry Swift, *Portrait of Imogen Cunningham,* 1934. Collection of the Imogen Cunningham Trust.
Back cover: Crawford Barton, *Imogen 1974,* 1974. Collection of Richard Lorenz, Berkeley, California. Copyright © 1993 Estate of Crawford Barton.

All reproduction photography is by Rondal Partridge, except as follows: Center for Creative Photography, figs. 28, 31, 32, 40; M. Lee Fatherree, figs. 25, 26; Richard Lorenz, fig. 18; The Oakland Museum, fig. 27.

Bulfinch Press is an imprint and trademark of Little, Brown and Company (Inc.)

PRINTED IN SINGAPORE

Contents

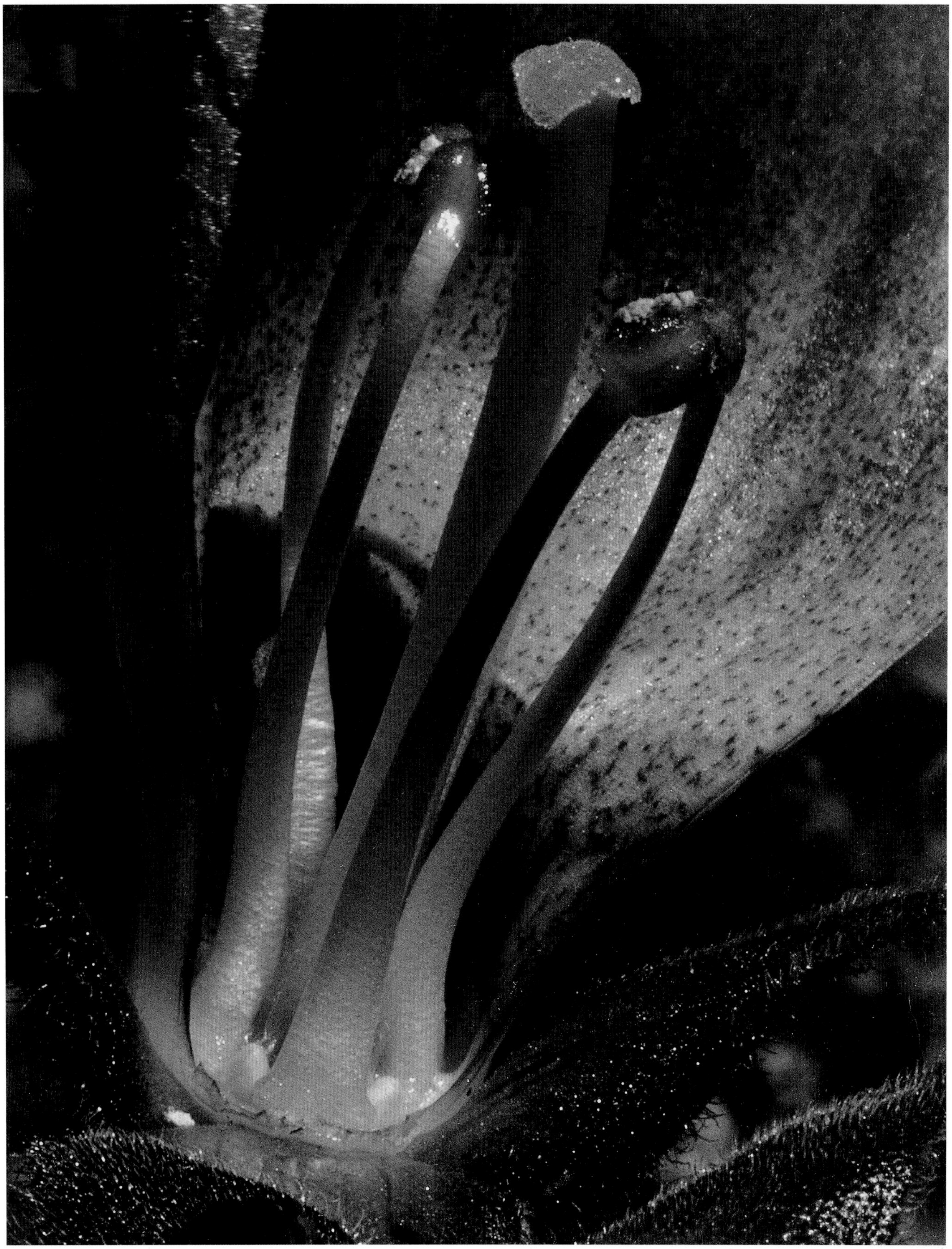

1. Imogen Cunningham: *Plant Form,* early 1920s

Imogen Cunningham: *Flora*

> To see a world in a grain of sand
> And a heaven in a wild flower,
> Hold infinity in the palm of your hand
> And eternity in an hour.
>
> —WILLIAM BLAKE, ca. 1803[1]

The paradox of expansion via reduction becomes vivid when one looks closely at the visual aspects of nature. The magnification of the precious detail of life forms almost guarantees provocative insight into the archetypal components of our world and its amazing, mystifying, and occasionally otherworldly formations (fig. 1). These microcosms, extracted from the broad context of our biosphere, can astound us with the uncanny plan of the whole order of nature.

The botanical photography of Imogen Cunningham is predicated on the reductive examination of diverse flora and the extraction of their essence, in split-second moments. One of the premier photographers of the twentieth century and perhaps America's quintessential woman photographer, Cunningham devoted some seventy years of her life to following the expansive scope of her artistic vision. Over her lengthy career, the style of her imagery evolved naturally and logically, closely paralleling many of the technical and aesthetic developments of the medium. She proved conclusively very early on that she had a flair for challenging, cogent observation, and that she could translate her vision into well-crafted, extraordinarily beautiful pictures. Her images are never formulaic but do categorically breathe of creativity, palpability, and integrity.

From the earliest years of her photographic career, which began about 1906, until her death in 1976, Cunningham explored botanical imagery with curiosity and determination. In the symbolist tableaux of her formative period, created in Seattle from 1910 until 1917, she often employed aspects of nature—from the intimate scale of a tabletop still life to the juxtaposition of the nude human form within the grandeur of the natural landscape—to suggest evocative, imaginary tales. Later, in California, she frequently replaced the external symbolism of her narrative images with the intrinsic symbolism of the vegetation she now photographed. Her youthful fascination with flora became a passionate amateur interest she would sustain throughout her life. As Emerson wrote: "The stars awaken a certain reverence, because, though always present, they are inaccessible; but all natural objects make a kindred impression, when the mind is open to their influence. . . . Nature never became a toy to a wise spirit. The flowers, the animals, the mountains, reflected the wisdom of his best hour, as much as they had delighted the simplicity of his childhood."[2]

Cunningham's photographs of plant, shrub, and tree forms might easily connote a pantheistic doctrine, but unlike Emerson, she did not extrapolate from nature the omnipresence of a universal being. More likely for Cunningham, image making simply celebrated the visual aesthetics, intricacies, and lucidity she found in nature and its basic, cyclical facts of life, death, growth, and regeneration. "Seeing is perhaps the greatest part in the education of a photographer," she stated. "Altho I have no definite formula for the seer, it does appear that effective seeing is to be acquired mainly by practice. What one sees can be of equal importance with how one sees it. . . . I do find myself, from time to time, lost in some tiny aspect of nature."[3] Integrating the forms and spirit of the flora populating our earthly horizons in her bold designs and eloquent compositions, Imogen Cunningham, with her technical skill and frontier wisdom, enriches our minds and delights our souls. Her vision in this particular field is matched by no other.

Cunningham's interest in botanical still life might have begun as early as 1909 during her studies at the Technische Hochschule in Dresden, Germany, where she took courses under the world-renowned photochemist Robert Luther. The 1909 International Photographic Exposition, one of the largest such displays ever attempted, was in progress in the city, providing Cunningham with a unique opportunity to view and compare the work of European and American photographers. Here she claimed to have seen "the best of American photography placed side by side with the foremost work of the European artists, and in technical skill, good taste, and real pictorial values the American work stood first."[4] Imogen later recalled: "It's strange that the person whose work I remember best at that time was Baron [Adolf] de Meyer, who later came to America and worked for the Condé Nast Company. At that time he lived in Europe and was doing all sorts of delicate and beautiful things with table decorations and fancy elegant women."[5] De Meyer focused on flowers, such as hydrangeas and water lilies arranged in translucent vessels, and washed his images with a soft luminescent light (fig. 2). The ethereal feeling and eccentric pictorialist space of his photographs must have been impressive and memorable to a young student photographer whose firsthand experience with fine photography had been limited to working in the Seattle studio of Edward S. Curtis, the chronicler of the North American Indian.

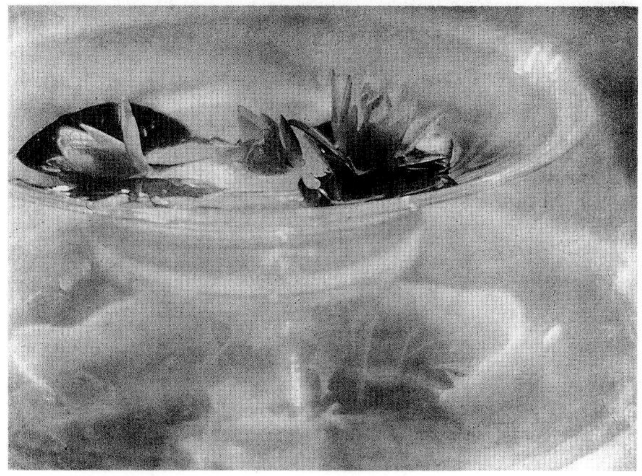

2. Adolf de Meyer: *Still Life,* 1908. Reproduced in *Camera Work,* no. 40 (1912).

Imogen's journey home to Seattle took her through New York City, where she visited Alfred Stieglitz's "291" gallery, the most influential source in America for modern art, photography, and aesthetic ideology. She also stopped in Washington, D.C., to visit a sister, and there she prophetically bought a magnolia, a plant that would eventually become her signature image: "I bought a *Magnolia grandiflora* at the Arlington cemetery. The stock was taken from a tree that George Washington had planted. . . . Mind you, all the way across the continent I paid someone on the train to water and take care of that plant. I remember it arrived in San Francisco in good condition and it went with me to my father's ranch in Sonoma County. It was in a gallon can, so it was a pretty sizeable tree I was carrying around. And if you knew how hard it was to get to Sonoma County from San Francisco! I liked this plant very much and I thought it would be wonderful on the ranch, but it died later because my father relocated it."[6]

Cunningham's earliest use of flora in her own photography began about 1910. She had returned to Seattle in September of that year and soon established a fledgling portrait studio. To personalize and authenticate the realism of her commercial portraiture, she began to include simple floral arrangements with her sitters. A fine example is her charming self-portrait with a bowl of flowering narcissus (pl. 1), her choice of plant perhaps a quaint metaphor for this self-mirroring study—an act of self-examination, confidence, and validation. Cunningham was skeptical of physical beauty, perhaps because she considered herself lacking in traditional prettiness, and she noted that "one must not have a too pronounced notion of what constitutes beauty in the external, and, above all, must not worship it. To worship beauty for its own sake is narrow and one surely cannot derive from it that aesthetic pleasure which comes from finding beauty in the commonest things."[7] Her statement comfortably applies not only to her portraiture but to *This Is My Garden,* her still life of a dish garden, carefully rendered in muted values of platinum (pl. 2). The plant form she offers, a weedlike, medicinal herb native to China and Japan, recalls a thought from Maurice Maeterlinck's contemporaneous essay on the virtues of weeds:

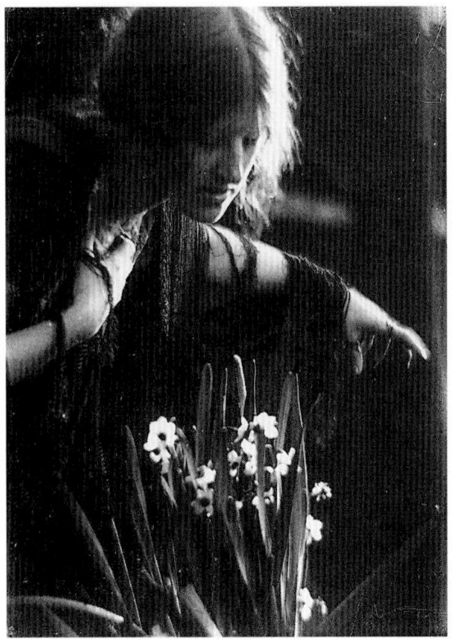

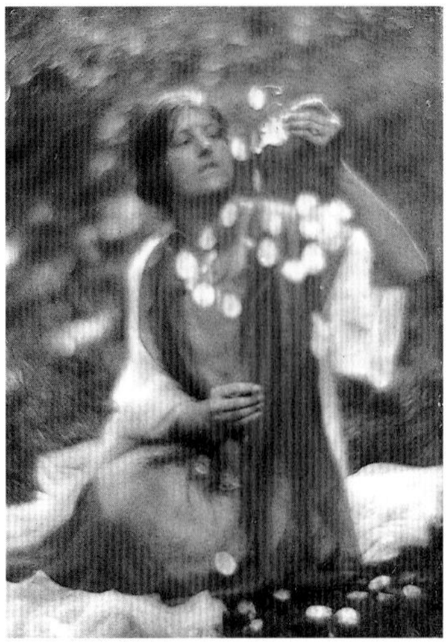

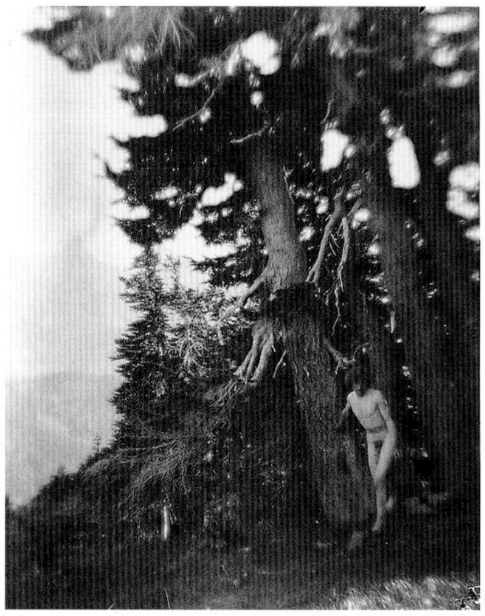

LEFT TO RIGHT:

3. Imogen Cunningham: *Clare with Narcissus,* 1910

4. Imogen Cunningham: *The Honesty Plant,* 1912

5. Imogen Cunningham: *On Mount Rainier 6,* 1915

> They keep the secret of a mysterious mission. They are the indelible primitives. The soil is theirs since its origin. They represent, in short, an essential smile, an invariable thought, an obstinate desire of the Earth. That is why it is well to question them. They have evidently something to tell us.[8]

Flora next became a crucial prop in Cunningham's creation of symbolist mythologies, in which she frequently included friends who shared an adjoining studio: John Butler, a mural painter; Clare Shepard, a painter of miniatures; and John's brother, Ben. The three became her models in romantic, mysterious tableaux (fig. 3). In period costume, Imogen's friends reenacted aspects of their favorite works of literature, such as William Morris's medieval tale, *The Wood Beyond the World* (1894), or the exotic verse of Japanese poet and art critic Yone Noguchi. Cunningham's increasing interest in botanica is apparent in many of her portraits of Clare; in one image she wove a metaphorical title and evocative scene around the common name of the plant species *Lunaria annua,* or honesty plant (fig. 4). Noguchi's poem, "The Garden of Truth," might have been her inspiration:

> A light-walled corridor in Truth's palace; a humanity-guarded
> chapel of God, where brave divinities kneel, small as
> mice, against the shoreless heavens,—the midnight
> garden, where my naked soul roams alone, under the
> guidance of Silence.[9]

From the time of her marriage to the artist Roi Partridge[10] in 1915, Cunningham seemed to broaden her concept of flora and nature. Roi preferred to sketch directly from nature and greatly appreciated the topography of the Pacific Northwest. He and Imogen made frequent trips into the Mount Rainier terrain, where each found stimuli for their artistic endeavors. Cunningham engaged Roi as a model in an extended series of the nude in the wilderness, juxtaposing his lithe body with mammoth hemlock tree trunks and the grandiosity of the mountains (fig. 5). With the arrival of their first son, Gryffyd, later that year, Imogen became a working mother, balancing household and professional commitments, but she still found time to document her young son's activities, such as his innocent wonderment at a handful of nasturtiums (fig. 6).

In 1917 the family moved to northern California, and Cunningham gave birth to twin sons. Imogen devoted the years 1917 to 1920 largely to caring for her boys. Although she had no studio or commercial clients, she never ceased to photograph her family and intimate surroundings in their San Francisco home and at her parents' Graton ranch. The images of this period are eloquent examples of naturalism—objective, atmospheric, loving yet unsentimental (figs. 7, 8). Reminiscent of the late-nineteenth-century photographs of Britain's P. H. Emerson, her pictures also reflect his viewpoint: "The nearer we get to Nature the sweeter will be our lives, and never shall we attain to the true secret of happiness until we identify ourselves as part of Nature." [11]

In the summer of 1920 the Partridges moved to the Fruitvale section of Oakland, a remote, rural area. Roi planted trees, and Imogen concentrated on gardening, especially promoting succulent growth on the property's rock walls. She remained creatively active during the early years of motherhood with, as she wrote, "one hand in the dishpan, the other in the darkroom." In a lecture note she later wrote in the third person, she described herself during this period as having "a skill with the camera, which she was not willing to sacrifice to maternity, so she turned her camera to use and photographed the things she had around her—her own children of course and plants that she cultivated. It is quite easy to do a bit of gardening work and yet attend children. It is not as easy to do good photographic work, but it can be done. She did both."[12]

Imogen and Roi began to meet other artists and photographers of the day in the Bay Area. In 1920 they befriended the photographer Edward Weston and his colleague Johan Hagemeyer. They offered their work to each

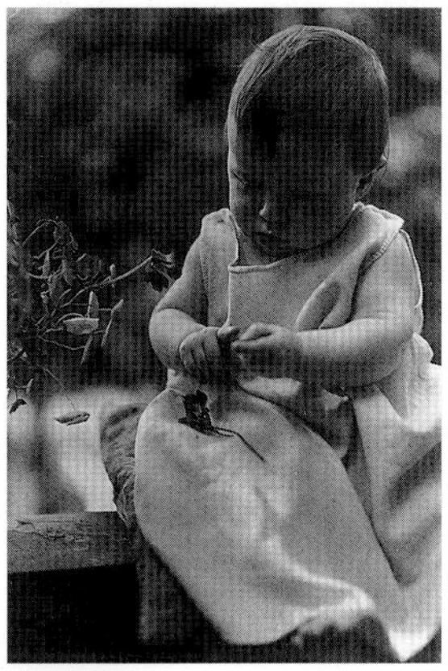

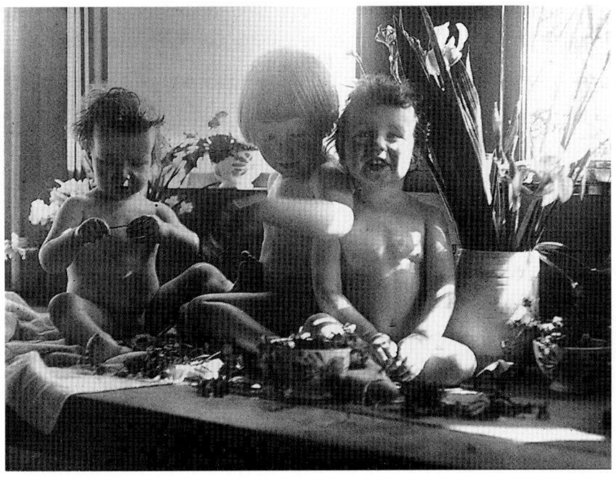

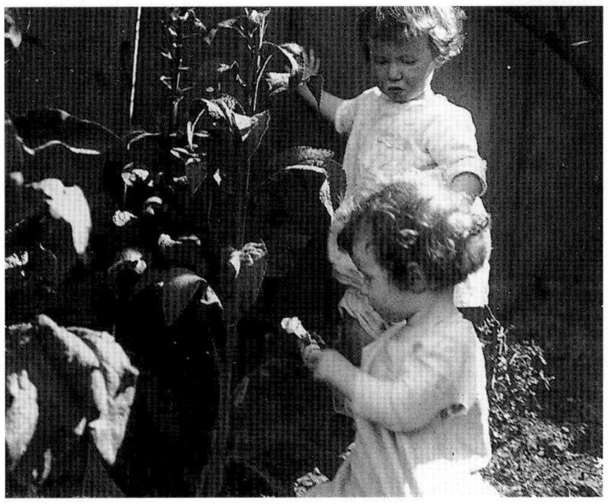

TOP:
6. Imogen Cunningham: *Gryff with Nasturtiums*, 1916

BOTTOM LEFT, RIGHT:
7. Imogen Cunningham: *Boys with Cut Flowers*, 1919
8. Imogen Cunningham: *Twins Picking Foxglove Buds*, 1919

other, as well as that of Margrethe Mather, Weston's studio partner, for viewing and discussion. Meeting a group of committed, artistic friends undoubtedly incited Cunningham to resume her professional work. By 1921 she had accepted her first commercial portrait session in five years, an occasion that might have rejuvenated the best of her creative energies, for 1921 marks a distinct turning point in Cunningham's work. She refined her approach to nature, switching her focus from long to near. She began to study detailed pattern and form, visible in her images of bark texture and twisted tree trunks along the Point Lobos cliffs (pl. 3, fig. 9), the succulent plant forms clinging to coastal rocks (fig. 10), and the glistening seaweed tossed onto the Carmel beach (fig. 11).

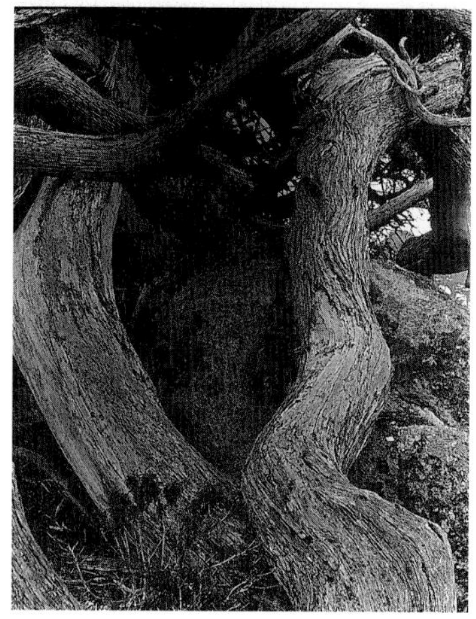

9. Imogen Cunningham: *At Point Lobos 2,* 1921

Cunningham empowered her images by isolating her subject; she minimized the background, expanded scale with close-up scrutiny, and formalized presentation. Essentially previsualizing her photographs, she drew attention to form by subtracting extraneous material and closely focusing her lens, frequently producing negative plates (especially during the 1920s and 1930s) of 1:1 proportion. An emphasis on clarity, form, and definition displaced her previous use of pictorialist space. Interestingly, Cunningham's homage or response to Mather's *Pine Branch* (fig. 12), an image she acquired in the early 1920s, reveals the difference. Mather, a remarkable aesthete who integrated elements of *japonisme* into her imagery (and into the pictorialist

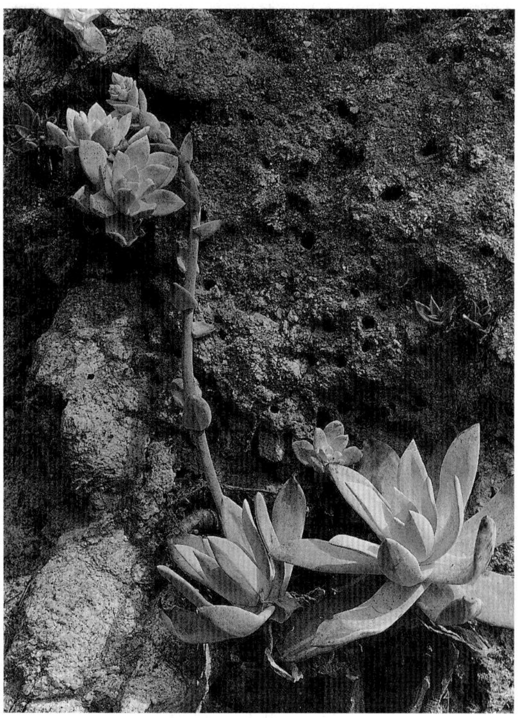

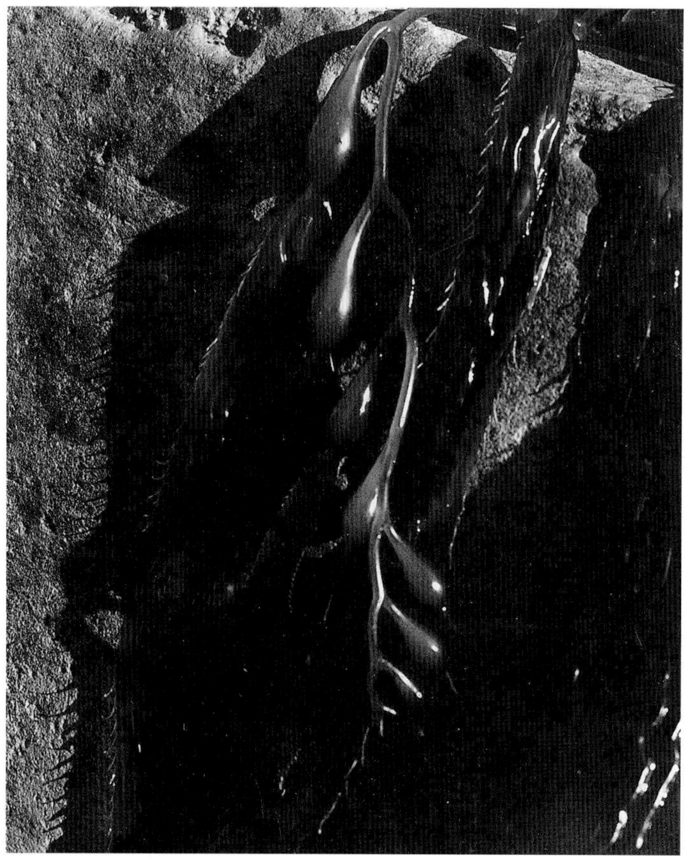

LEFT:
10. Imogen Cunningham: *Succulents, Point Lobos,* 1921

RIGHT:
11. Imogen Cunningham: *Seaweed 2, Carmel,* 1921

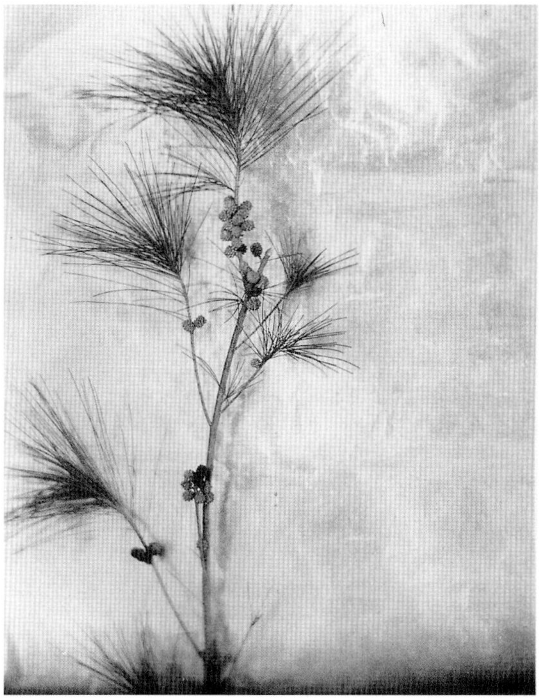

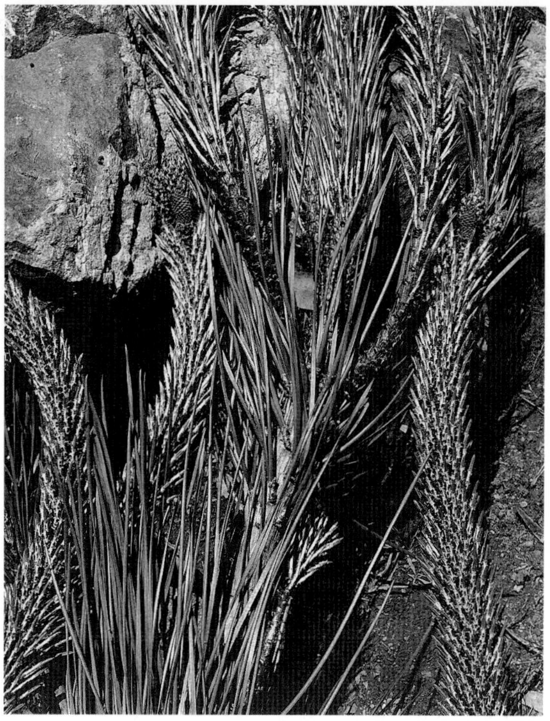

work of Edward Weston as well), created a spare, strikingly handsome design with a simple branch. As in a Hiroshige print, negative space is as critical to the composition as the design elements. Cunningham's new approach (contemporaneously duplicated by Mather) gets to the point and excises the empty space. Her own pine tree study (fig. 13) is a crisp, scrutinizing detail meant to illuminate natural design rather than create an artistic decoy.

As she juggled family life, commercial sittings, and creative work, Imogen turned particularly to flora, selecting choice specimens from her garden and surrounding areas. She noted: "The reason I really turned to plants was because I couldn't get out of my own backyard when my children were small. That was when I started photographing what I had in my garden; it was mostly succulents, which I called cacti. When I learned that all succulents were not cacti, but that all cacti were succulents, I decided to learn a little. I've been working at it ever since."[13] Cunningham's fascination with botanical imagery had early antecedents: she made botanical lantern slides for a college professor and had photographed her Japanese dish garden before 1915 (pl. 2). Her interest in flowers is likewise peripherally apparent in her commercial and artistic works. But her favorite magazine, *Vanity Fair*, had published in recent issues two classic floral still lifes, perhaps influencing her to change her approach to botanical subjects.[14] Charles Sheeler's mannered, proto-precisionist tabletop still life, *Zinnia and Nasturtium Leaves*, was published in May 1920, and Edward Steichen's *Lotus, Mount Kisco, New York*, a resplendent study of a sun-dappled flower floating among rhythmic leaf forms, was reproduced in July 1923. The formal arrangement and clear, close-up scrutiny

LEFT:
12. Margrethe Mather: *Pine Branch*, ca. 1918. Platinum print. The Metropolitan Museum of Art, New York (ex-collection Imogen Cunningham).

RIGHT:
13. Imogen Cunningham: *Pine Tree*, early 1920s

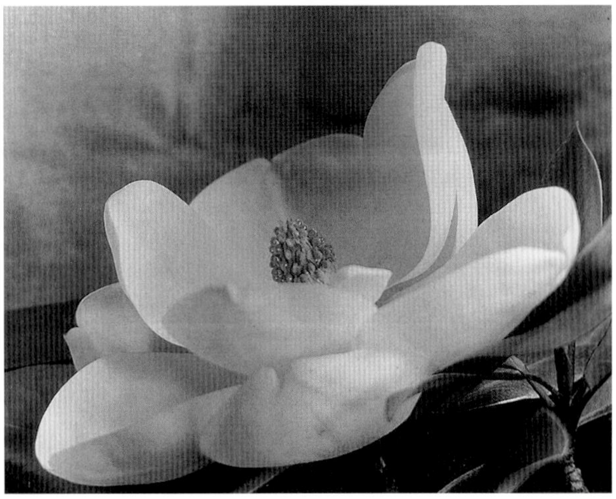

of each picture were a challenging rebuke to conventional pictorialism. In a similar vein, from 1923 to 1925 Cunningham made an extended study of magnolia flowers (fig. 14) which grew more simplified as she emphasized the blossom's essential form. Two of her signature images come from this series, *Magnolia Blossom* (pl. 11) and its variant detail, *Magnolia Blossom, Tower of Jewels* (pl. 12), a closer examination of the flower's core.[15]

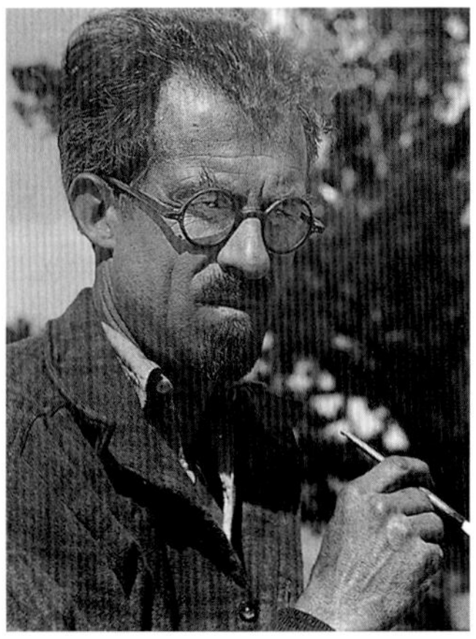

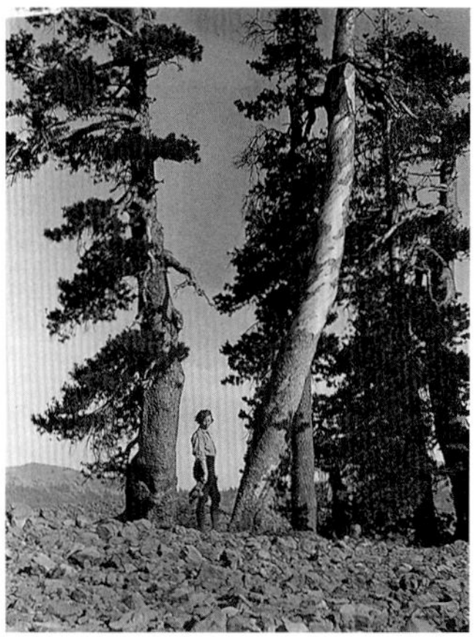

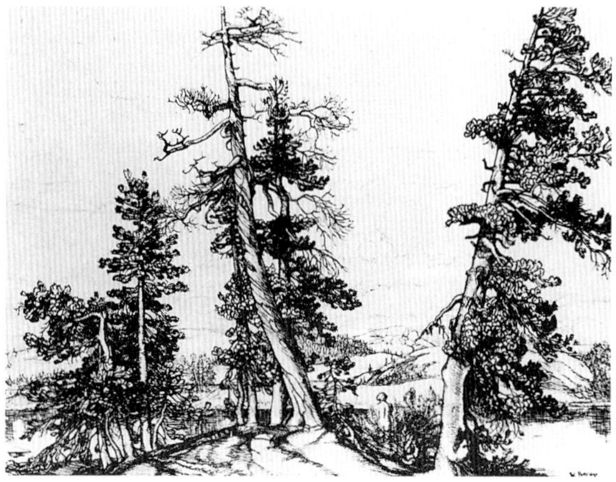

The Partridge family made frequent outings into the California landscape. Summer meant an extended camping trip for the boys, which netted some artistic diversions for Imogen and Roi. A love and appreciation for nature went hand in hand with their travels, and their art became a partial travelogue. During 1925 they motored to the Sierra, stopping at Summit Lake on Donner Pass. Comparable to the evolving refinement of her viewpoint of flora, Cunningham's portrait of Partridge (fig. 15) is condensed and tight, whereas Roi's photograph of Imogen (fig. 16) is a more expansive formal landscape that frames his wife between rocks, sky, and trees. Cunningham's spectacular photograph of a magnificent aged juniper tree (pl. 6), taken on the same trip, points to the veracity of her new vision and her emotional connection to natural forms. Partridge's etching of a nude Imogen at Summit Lake captures the spirit of her association with nature (fig. 17).

TOP:
14. Imogen Cunningham: *The First Magnolia,* ca. 1923

BOTTOM, LEFT TO RIGHT:
15. Imogen Cunningham: *Roi Partridge at Donner Pass,* 1925
16. Roi Partridge: *Imogen Cunningham at Donner Pass,* 1925. The Imogen Cunningham Trust, Berkeley, California.
17. Roi Partridge: *Mountain Lake,* 1925–28. Etching on copper. Courtesy of Anthony R. White, San Mateo, California.

Cunningham's botanical interests were validated by some of her photographer-friends. The Dutch photographer Johan Hagemeyer, who had been a professional horticulturist in Amsterdam, Berkeley, and southern California,[16] created a limited body of floral still lifes during the 1920s in which roses, dahlias, tulips, calla lilies, and magnolias (fig. 18) all exhibit a romantic chiaroscuro. Margrethe Mather also made subtle, fluidly arranged studies of plant life during the 1920s (fig. 19).

In the late 1920s Cunningham met Egon von Ratibor, a German prince who more commonly called himself James West: James because he liked it, and West because he had settled in the American West.[17] West had participated in the creation of the botanical garden at the University of California at Berkeley, and Cunningham continued her botanical education in his company. "I had taken some plant pictures before then [meeting West]," she said, "but I didn't really know anything about the material. I called all succulents 'cactus,' and that sort of thing."[18] West introduced her to the California Horticultural Society, of which he was a founding member, and together they sought out unusual plant subjects, particularly succulents.

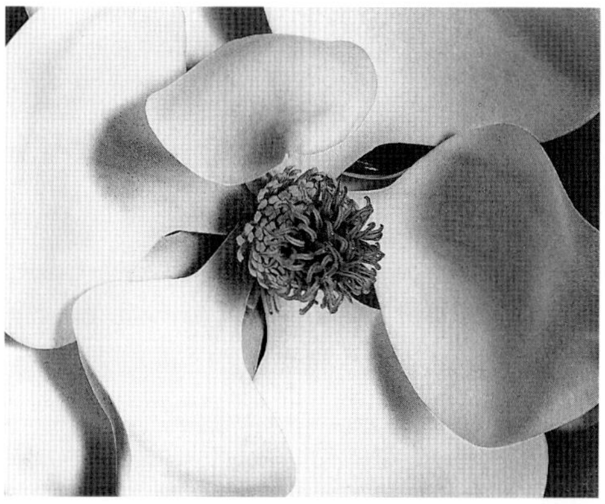

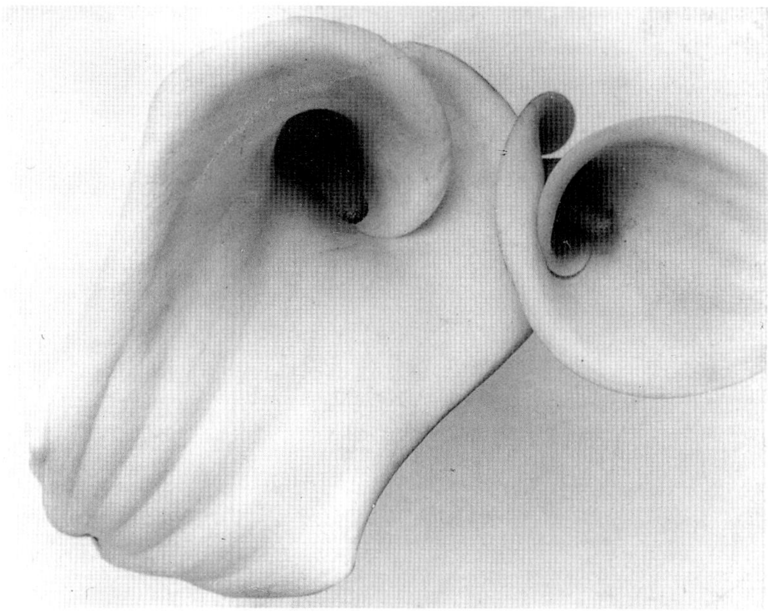

TOP:
18. Johan Hagemeyer: *Magnolia,* 1929. Center for Creative Photography, University of Arizona, Tucson.

BOTTOM:
19. Margrethe Mather: *Calla Lilies,* ca. 1926. Estate of William Justema, Berkeley, California.

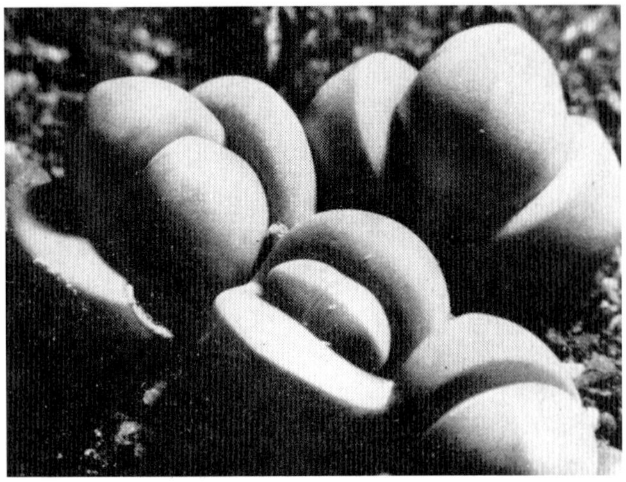

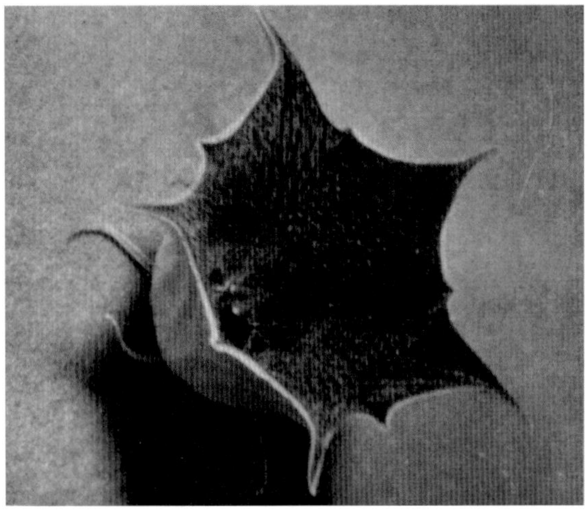

West had come to the United States in 1915. He became stranded by the Allied naval blockade of World War I, and, cut off from his funds, he took a job in a lumber camp near Fort Bragg, California. He moved to Arizona and worked as a gardener and landscape horticulturist, developing great expertise in cacti and succulents. During the early 1930s, West frequently lectured to horticultural groups in the San Francisco Bay Area and was an associate editor and frequent contributor to the *Journal of the Cactus and Succulent Society of America.* An eccentric who lived in a tent in the backyard of a boarding house, West nevertheless was considered genteel, brilliant, and sophisticated. He occasionally stayed at the Partridge home, where a guest bed was always available to him.

West was an accomplished botanical draftsman as well as a competent photographer. His writing, too, speaks well of the breadth of his interests and his knowledge of art history. In a 1929 article on the plant form *Mesembryanthemum,* he commented on his own photograph (fig. 20):

> The accompanying illustration might pardonably be mistaken for a conception of abstract form, chiseled out of marble by Brancusi or some other modernist. The strange objects represent, however, only another one of the many remarkable adaptive growth-forms among the Mesembrieae. It would be difficult to conceive succulence in a leaf, reduction of surface in relation to volume, carried much farther than here. The surface is smoother than an eggshell, of a white faintly tinged with gray-green. Striking as is the appearance of the plants, it may easily be imagined how inconspicuous they would be among the white quartz rock which is their natural habitat, another instance of protective mimicry in the family.[19]

Another brilliant example of West's word and image is a 3x magnification of a stapelia (fig. 21), a plant with "a peculiar and unholy attraction," which he compares to "a bridal bouquet for His Infernal Majesty . . . with the scent of corruption about them, their marriage-bed a haven for carrion-flies."[20]

As teacher and friend, West was a significant influence on Cunningham, particularly in her occasional taste for bizarre succulent forms. For example,

LEFT:
20. James West: *Seedlings of Argyroderma Testiculare,* ca. 1929. Reproduced in *Journal of the Cactus and Succulent Society of America,* vol. 1, no. 3 (1929).

RIGHT:
21. James West: *Flower of Huernia Penzigii,* ca. 1929. Reproduced in *Journal of the Cactus and Succulent Society of America,* vol. 1, no. 5 (1929).

her dramatic close-up of a *Faucaria tigrina* (fig. 22), commonly known as the "Tiger Jawed," presents the succulent like the gaping mouth of a long-toothed creature. Such imagery steered Cunningham's pictures away from the picturesque and probably were part of a conscious strategy to eschew conventionally pretty, romantic subjects. Cunningham was a sophisticated aesthete and often delighted in upsetting common values. "I am always interested in anything that is exotic," she declared,[21] and it seems no surprise that she, in fact, never photographed a rose.

Cunningham did not always photograph plant materials straightforwardly or in their natural habitats. She collected specimens from her yard, the neighbors, and surrounding fields; arranged them in rhythmic compositions against velvet backdrops; and shot under a combination of daylight and blue tungsten photofloods, as in her many studies of calla lilies (pls. 8, 9, 10). Sometimes the images were created spontaneously in a spirit of fun, albeit with a solid, intuitive sense of design, as in her many manifestations of neocubist patterns employing the central sheaths of an *Agave americana* (pls. 19, 20).

Whereas Edward Weston defined a part of his oeuvre by his preoccupation with the quintessential life force of fruits and vegetables, Cunningham made only a few attempts with pears, bananas, pineapples (fig. 23), and erotically suggestive carrots and radishes. Whatever she photographed, however, transcended its reality through her inimitable style. As her acquaintance Amédée Ozenfant wrote: "All thanks to Daguerre; he revealed clearly that exact imitation, item for item, cannot reproduce nature, but that only EQUIVALENTS can. It is not by making volcanoes that sulphur is produced. Retorts are better. Nature is not reconstituted by imitating animal cries or natural noises, but by writing pastoral symphonies. To imitate something is merely to stuff it."[22]

After her studies in Dresden, Cunningham maintained ties to German culture by reading publications such as *Das Deutsche Lichtbild* (The German

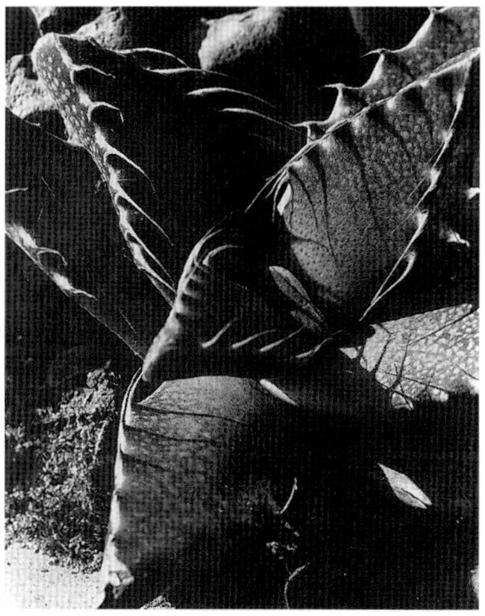

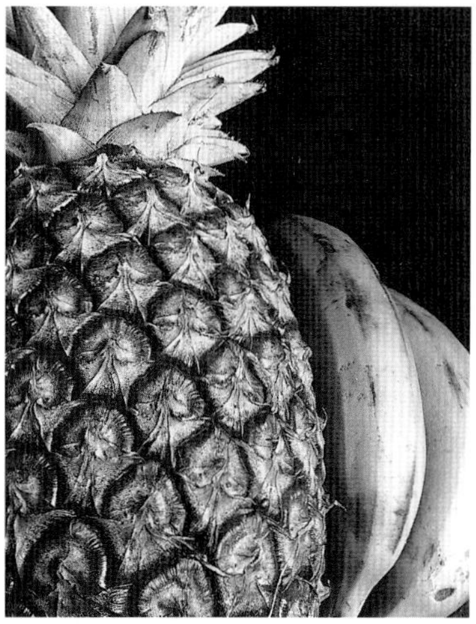

LEFT:
22. Imogen Cunningham: *Faucaria Tigrina,* 1920s

RIGHT:
23. Imogen Cunningham: *Still Life, Pineapple and Bananas,* 1920s

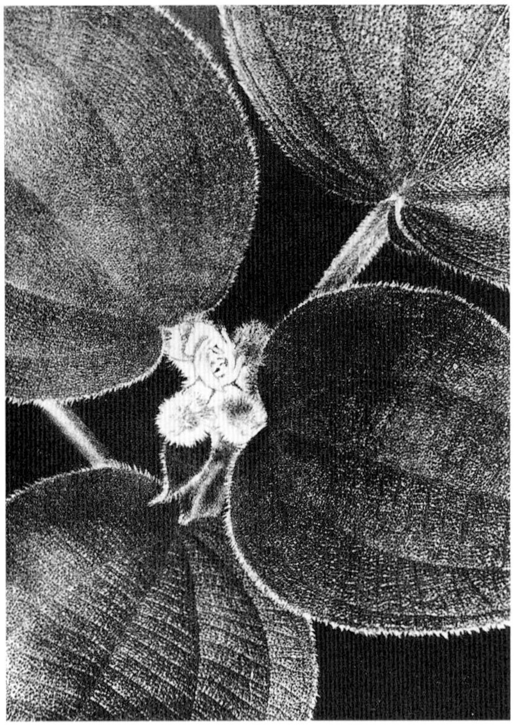 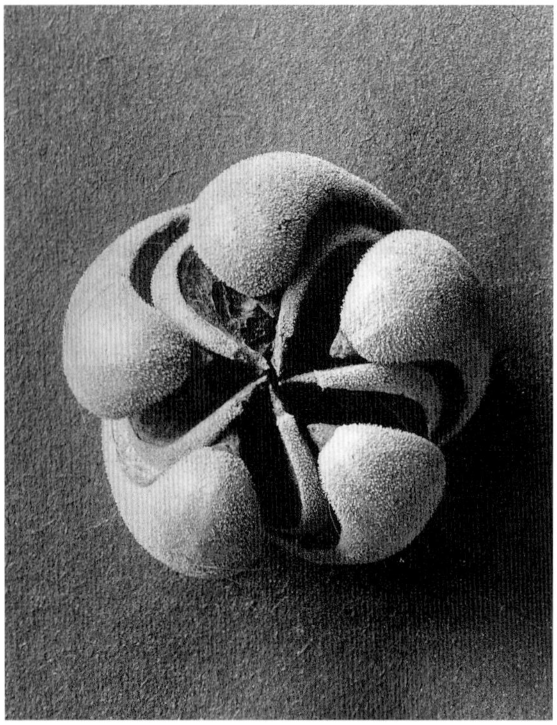

LEFT:

24. Albert Renger-Patzsch: *Heterotrichum Macrodum,* before 1928. Reproduced in *foto-auge* (Stuttgart: Akademischer Verlag Dr. Fritz Wedekind, 1929).

RIGHT:

25. Karl Blossfeldt: *Blumenbachia Hieronymii (Loasaceae),* before 1932. Collection of Richard Lorenz, Berkeley, California.

Photograph), which reproduced botanical studies by Albert Renger-Patzsch, Ernst Fuhrmann's *Die Welt der Pflanze* (The World of Plants, 1929–31), and Karl Blossfeldt's *Urformen der Kunst* (Art Forms in Nature, 1928). That Cunningham and Renger-Patzsch made comparable plant images is more likely due to parallel interests than direct influence, since Imogen was well into her own studies before becoming aware of Renger-Patzsch's work. In fact, one of the fundamental attributes of modernism, the perplexities inherent in scale, applied to organic forms, were explored by a number of photographers throughout Europe and America—Edward Steichen in New York, Paul Strand in Maine, Man Ray and Florence Henri in Paris, Jaromír Funke in Prague, Emili Godes in Barcelona, and Karl Blossfeldt and Renger-Patzsch in Germany.

Renger-Patzsch was closely associated with *Die Neue Sachlichkeit* (New Objectivity), a modernist German style noted for its clear and objective presentation of visual information. His formal examinations of specific flora (fig. 24) record intrinsic natural designs and create analogies between elemental botanical structures and aspects of industrial design. Believing that the camera was better able to record natural objects than the eye, Renger-Patzsch typically selected a characteristic fragment from the whole, hoping to underscore its essential elements through isolation.[23] The influential Bauhaus teacher László Moholy-Nagy also considered the camera capable of "seeing" what the eye could not: "In photography we possess an extraordinary instrument for reproduction. But photography is much more than that. Today it is in a fair way to bringing (optically) something entirely new into the world."[24]

Karl Blossfeldt photographed plant forms from about 1900 to 1930 for his course lectures on "Modeling from Plants" which he taught at the Kunstgewerbeschule in Berlin. At this school of applied arts, the study of plant forms provided motifs for use in ornamental design as well as demonstrated that good industrial design often had been anticipated in natural forms. In simple, technical examinations (often under magnification) of hundreds of flora (fig. 25), Blossfeldt developed a unique style that aesthetically and theoretically was compatible with the objective "new vision" of colleagues like Renger-Patzsch. In the foreword to his second compendium of "Art Forms in Nature," Blossfeldt wrote:

> The plant may be described as an architectural structure, shaped and designed ornamentally and objectively. Compelled in its fight for existence to build in a purposeful manner, it constructs the necessary and practical units for its advancement, governed by the laws familiar to every architect, and combines practicability and expediency in the highest form of art. Not only, then, in the world of art, but equally in the realm of science, Nature is our best teacher.[25]

The philosopher Walter Benjamin noted that Blossfeldt's photographs "reveal a completely unexpected wealth of analogies and forms in the plant world. Only photography is able to do this, since it requires powerful enlargement before the forms lift the veil which our laziness has thrown over them."[26]

Cunningham's intuitive approach was not as austere, morphologically clinical, or encyclopedic as Blossfeldt's studies. Cunningham, however, probably more than any other West Coast photographer, paralleled the objectivity of the Germans in her work. Unlike Weston and his innate transcendentalism, her tendency toward plain descriptiveness allowed her to produce superior examples of unsentimental botanical imagery. Cunningham perhaps considered the paradox in her pictures; even though objectively rendered, they were produced by sensation and emotion:

> I have never thought of photography in connection with philosophy but if you want me to think of it as the dictionary puts it, as "knowledge of the causes of mind and matter," I will try. We can worry our heads over the reasons for photography just as we can worry them as to the reasons for ART. While photography is in the main, a representational medium, it can also be expressive—that is, it may show the skilled objectivation of the doer—be a free activity as is painting. The fact that it does not always use the language of feeling and set up an aesthetic transmission is the fault of the person behind the camera. But I would not know how to define that fault or failure.[27]

Cunningham's work of the late 1920s presents a strong case for recognizing her as the most sophisticated and experimental photographer on the West Coast. Commentators nevertheless often cite the notion that Edward

Weston was her modernist mentor. Cunningham herself refuted this idea; in a letter to San Francisco art critic Alfred Frankenstein, she noted that "all the plant things I did from 1925 thru 1933 or 34 were BEFORE Edward became interested in the things he produced later in Carmel, such as the pepper and the artichoke cut in half. By this I am not implying that he was influenced by me, but that I was NOT influenced by him."[28] A 1929 review of a photography exhibition at the Berkeley Art Museum declared: "Imogen Cunningham . . . is easily the star of the show. She has struck a beautiful balance between quality and organization; she has simplified, yet she has not simplified to the point where her work loses interest. . . . On the other hand, we see here the work of one who is probably the most exalted of all pictorial photographers: Edward Weston . . . yet, unlike Imogen Cunningham, Weston has simplified his compositions to a degree that results in practically no composition at all. There are not enough elements involved to make his works vitally interesting. Were it not for Cunningham's revelations of what can be created in photography, we might appreciate Weston the more."[29]

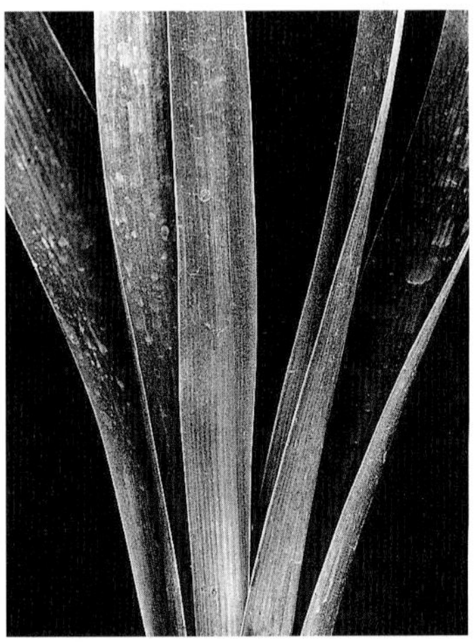

26. Brett Weston: *Lily Stalk,* 1925. Collection of Scott Nichols, San Francisco, California. Copyright Estate of Brett Weston.

In 1929 Weston was asked to select the West Coast entries for the exhibition *Film und Foto,* to be held in Stuttgart. This historic exposition largely defined the nature of avant-garde photography at the time. Cunningham was one of Weston's choices, for he found her work to be "fine and strong and honest."[30] After viewing her print of a false hellebore (pl. 22) at the Los Angeles County Museum, Weston wrote her: "I had one thrill and it was your print—Glacial Lily—it stopped me at once, I did not note the signature until I had exclaimed to myself—'this is fine!' It is the best thing in the show, Imogen, and if you keep up to that standard you will be one of a handful of important photographers in America—or anywhere. Thank you for giving me rare pleasure."[31] He was again generous with praise when he reviewed an exhibition of her work at a Carmel gallery two years later: "She uses her medium, photography, with honesty,—no tricks, no evasion: a clean cut presentation of the thing itself, the life of whatever is seen through her lens,—that life within the obvious external form. With unmistakable joy in her work, with the unclouded eyes of a real photographer, knowing what can, and cannot, be done with her medium, she never resorts to technical stunts, nor labels herself a would-be third-rate painter. Imogen Cunningham is a photographer! A rarely fine one."[32]

Although Weston asked Cunningham only for examples of her flower forms for *Film und Foto,* the ten photographs she exhibited in Stuttgart included, along with eight "Pflanzenforms" (pls. 8, 10, 11, 22, 24, 27, 37, 42), a male and female nude study and a precisionist image of an industrial water tank.[33] She always claimed interest in photographing everything that could be exposed to light, so her inclusion of a nude and an industrial study were obvious declarations of her visual eclecticism. Weston also invited his son,

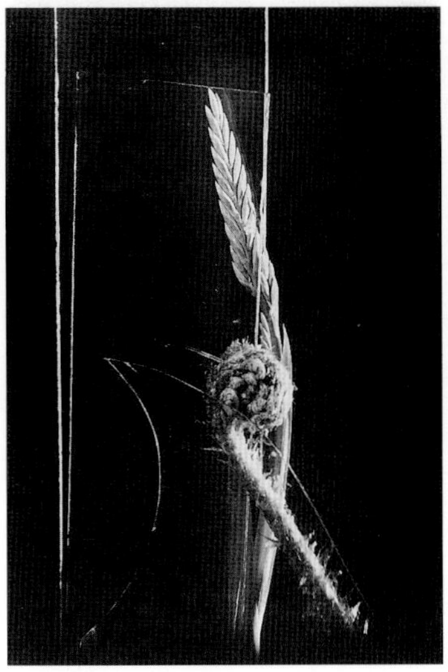

27. Roger Sturtevant: *Spring #3 (Pflanzenstudie mit Glas)*, ca. 1926. The Oakland Museum, gift of the artist.

Brett, and Roger Sturtevant, a San Francisco photographer, to submit work. Among a variety of subjects, each showed a plant form. Brett Weston's *Lily Stalk* (fig. 26) is a classic example of the new style of realism developing in California, while Sturtevant's photograph of layered plant forms and glass shards (fig. 27) seems rooted in a more experimental European vein. Edward Weston exhibited twenty of his own images including the flamelike *Cypress, Point Lobos* (fig. 28), the first photograph he took in the nature preserve that would consume his artistic interest for much of his life.

Cunningham discussed *Film und Foto* and her increasing participation in international expositions in a letter to a former Mills College friend:

> For fun I have turned to the unseen spots of nature—innards of flowers—and curious forms in all things. . . . For the last two years I have had quite some notice in German International exhibits, that in Stuttgart in '29 and in Munich in '30. A friend who did not know that my stuff was in Stuttgart happened upon it and wrote me that it was hung in a room with Steiner, Sheeler, Bruehl and Steichen. I am telling you this just to give you a clue to my style tho I might have been peeved if I had seen the room. Who can tell.[34]

A Cunningham retrospective of sixty-three prints opened at the M. H. de Young Memorial Museum on December 18, 1931. One critic noted that for "those who still think that certain forms of so-called modernistic art have no kinship to nature have but to see such direct photographic studies . . . to learn that a literal representation of nature may come very near to being an abstraction, for nature itself is capable of a more weird creative imagination than is the mind of man, 'modernist' or otherwise."[35] But another reviewer, a young amateur photographer named Ansel E. Adams, perceived in the work "a definitely feminine point of view" and criticized Cunningham for holding "a naive devotion to certain traditional conceptions of art. . . . Her prints could have been produced only by a woman, which does not imply, however, a lack of vigor. All her photographs brim with a restrained strength typical of keen decisive feminine energy."[36] Perhaps such criticism motivated Cunningham to select more severe, exotic, and even dangerous-looking botanical subjects for her photographs. To have her work accepted by her male peers, she seemed to masculinize it and augment the brute force of an image by using bold cropping and complementary background values (e.g., pls. 49–54). Alfred Stieglitz, America's arbiter of early-twentieth-century modern art and photography, in surveying a selection of her work in 1934, regarded *Calla Leaves* (pl. 54) as her finest photograph.[37]

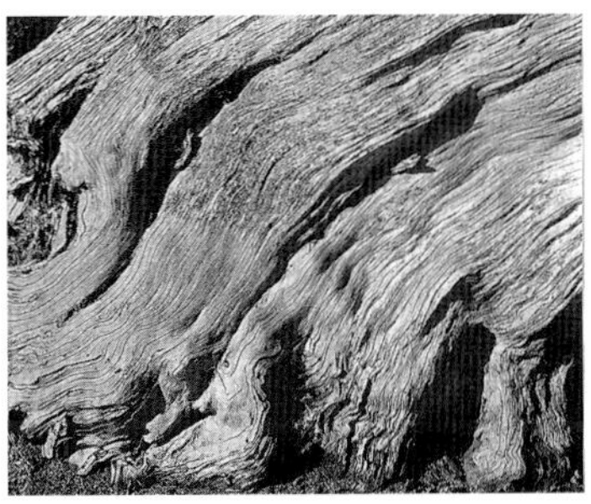

28. Edward Weston: *Cypress, Point Lobos*, 1929. Copyright 1981 Center for Creative Photography, Arizona Board of Regents, Tucson.

Another means of denying the female gender of the image's maker was to insinuate sexuality in the picture. In the early twentieth century, genital symbolism generally was employed only by men. Most notably within the context of European surrealism, eroticism in art and photography was created predominantly by and oriented toward the heterosexual male. Cunningham's suggestive images of unopened magnolia buds (pl. 40), the unraveling bract of a calla lily, tumescent black lily shafts (pls. 41, 42), and the labial leaves of a water hyacinth (pl. 44) are all charged with a natural erotic tension. Like Man Ray's depiction of a sensually enigmatic peach (fig. 29), Cunningham's shot of a sprouting avocado pit (fig. 30) is sexually provocative yet crosses over into a didactic presentation of intrinsic life forces. Liberal and open-minded, Cunningham was compulsively intrigued by life, wonders of nature, and reproductive functions, and although no direct evidence supports an erotic interpretation of her botanical imagery, their implied, occasionally risqué connotations seem obvious and more than subliminal. Man Ray described his ideas about the artist and primal eros in his 1934 essay, "The Age of Light":

> An effort impelled by desire must also have an automatic or subconscious energy to aid its realization. The reserves of this energy within us are limitless if we will draw on them without a sense of shame or of propriety. Like the scientist who is merely a prestidigitator manipulating the abundant phenomena of nature and profiting by every so called hazard or law, the creator dealing in human values allows the subconscious forces to filter through him, colored by his own selectivity, which is universal human desire, and exposes to the light motives and instincts long repressed, which should form the basis of a confident fraternity.[38]

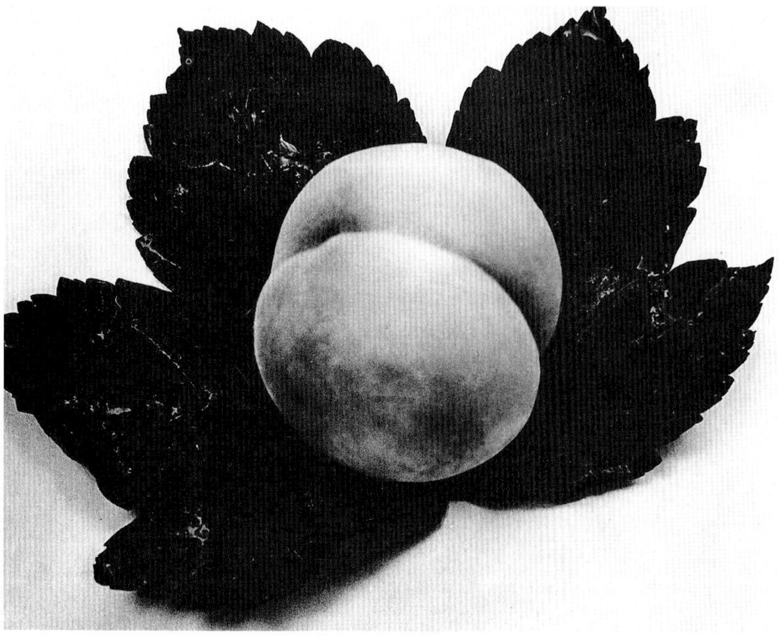

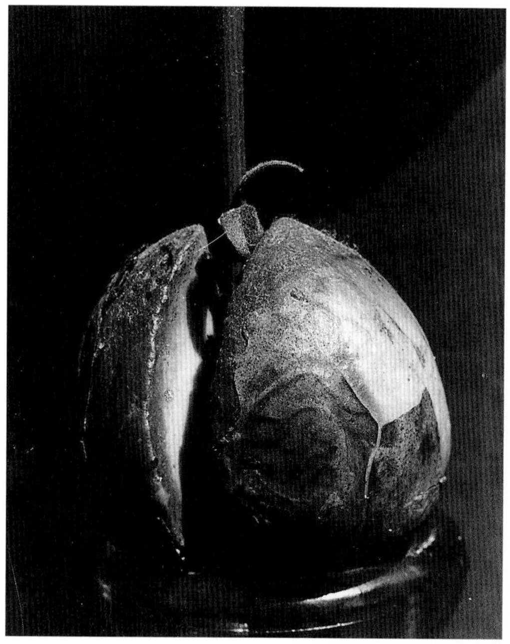

LEFT:
29. Man Ray: *Untitled*, 1931. Reproduced in *Photographs by Man Ray: 1920 Paris 1934* (1934).

RIGHT:
30. Imogen Cunningham: *Avocado Pit*, 1920s

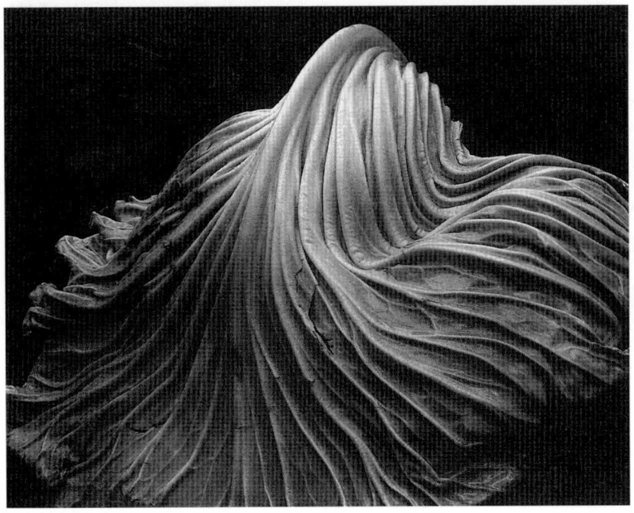

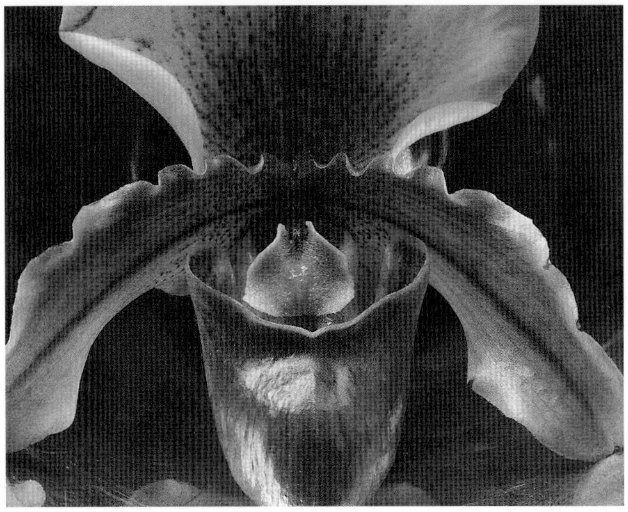

LEFT:
31. Edward Weston: *Cabbage Leaf,* 1931. Copyright 1981 Center for Creative Photography, Arizona Board of Regents, Tucson.

RIGHT:
32. Edward Weston: *Orchid,* 1931. Copyright 1981 Center for Creative Photography, Arizona Board of Regents, Tucson.

Edward Weston surprisingly denied the cool eroticism of many of his organic still lifes. In discussing one of his studies of glistening, undulating green peppers, he claimed that his photographs took "one into an inner reality,—the absolute,—with a clear understanding, a mystic revealment."[39] His goal was not to celebrate the inherent sensuality of plants but to abstract their form. Writing about a variant image from his cabbage series (fig. 31), he said: "This time it was a single leaf I used, achieving the strongest, most abstract results. . . . The power in that cabbage leaf is in mind as I write this."[40] Weston seemed to find fault with too much perfection, and, after having photographed an orchid given to him by a friend of Cunningham's (fig. 32), declared it as "too perfect—like trying to photograph a cathedral,"[41] perhaps one reason that he photographed so few flowers.

In his landmark book *Die Welt Ist Schön* (The World Is Beautiful), Albert Renger-Patzsch heralded the freeing of photography from the lexicon of traditional art: "The secret of a good photograph, one that possesses esthetic quality of a work of art, lies in its realism. . . . Let us leave art to the artists and let us try by means of photography to create photographs which can stand alone because of their *photographic* quality—without borrowing from art."[42] Carl G. Heise, Renger-Patzsch's editor, also emphasized photography's intrinsic nature: "The translation of what the photographer saw in everyday reality into a black-and-white image spurs the viewer on to seek a visual excitement which he previously passed by unaware."[43] Norman Bel Geddes, who reproduced and discussed Cunningham's *Magnolia Blossom* (pl. 11) in his landmark book on modern design, *Horizons* (1932), also validated photography as an art form: "How or by what means an artist expresses himself is of no consequence. It is the vitality and subtlety of feeling which results that gives the work its place in the sun, and it makes no difference whether it was achieved with paint, stone or machine. Camera still life illustrates emphatically that interest in visual form is not the result of subject matter. . . . Has it ever occurred to you that a photograph of a flower, even though devoid of color, might be as thrilling as a painting of it."[44]

In late 1932 several Bay Area photographers sharing a mutual ideology formed the Group f.64. Derived from the smallest aperture available on a large-format camera, the name implies images of the greatest depth of focus and sharpest detail. The founding members were Ansel E. Adams, Imogen Cunningham, John Paul Edwards, Sonya Noskowiak, Henry Swift, Willard Van Dyke, and Edward Weston. They sought to produce unmanipulated, sharply detailed photographs that were contact-printed without any retouching. The group and its assertion of visual reality became a benchmark for straight photography.[45]

In Group f.64's only exhibition at the de Young Museum, in 1932, Cunningham showed nine photographs, including seven botanicals (pls. 24, 37, 39, 43, 44, 45, 46). The other members exhibited an eclectic range of portraits, industrial and botanical still lifes, landscapes, and architectural studies—all seen with clarity and conviction. Group f.64, despite its short existence, has become surprisingly legendary considering that its mission was somewhat reactionary in its exclusion of the experimental techniques and styles often used by notable photographers of the European avant-garde. In fact, *Film und Foto,* three years prior, had promoted the various possibilities in contemporary applied photography: "the reality-photo, the photogram, photomontage, photo with etching or painting, and photos in connection with typography."[46] In the exhibition's companion publication, *photo-eye,* Franz Roh noted that innovative photographers "not only mean to say 'the world is beautiful,' but also: the world is exciting, cruel and weird."[47] Cunningham was the only Group f.64 member to have developed this all-inclusive attitude.

Group f.64 was only a facet of Cunningham's work, and its mandates did not necessarily guide or restrict her photography. Her interests were always too broad, her approach too flexible, to be constricted by someone else's definitions. Her innate curiosity led her to experiment, to "fool around," as she would often say. She did not hesitate to crop an image, save a print that had been accidentally solarized, sandwich negatives, or shoot double exposures. Elements of dada and surrealism touch her work, and the concept of the "pure" print with which she momentarily had associated herself was no more than one of many aesthetic considerations she visited.

In the early 1930s Cunningham increasingly explored the multiple image or double exposure. Other avant-garde photographers had stepped up their use of this pictorial device in the late 1920s, and Cunningham possibly was encouraged to experiment. She incorporated into her magazine work double exposures that contrasted expression and form, time and space. In an otherwise simple still-life arrangement of a *Protea cynaroides,* she manipulated one monumental blossom to magically create a whole bouquet by shifting

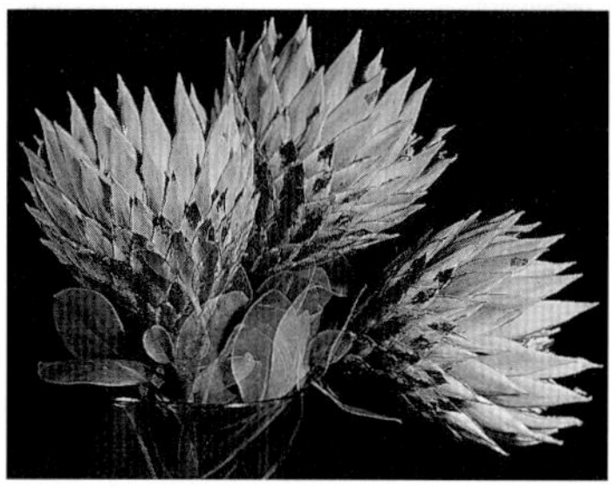

33. Imogen Cunningham: *Three Protea,* 1935

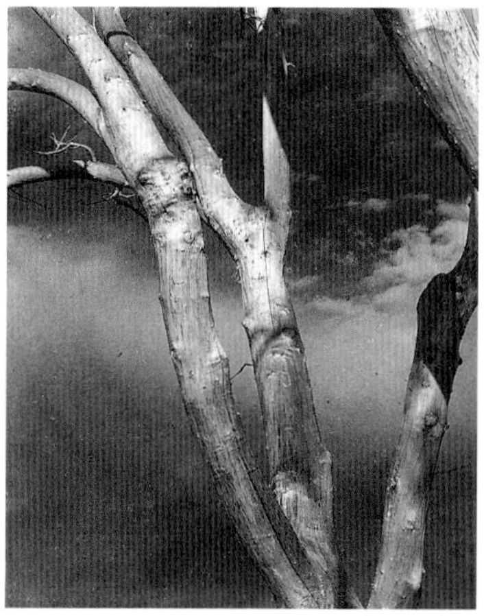

34. Imogen Cunningham: *Tree in the Sierra 1,* 1932

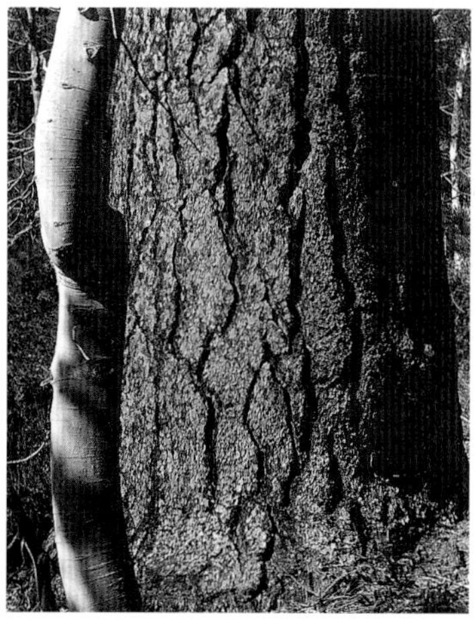

35. Imogen Cunningham: *Tree in the Sierra 2,* 1932

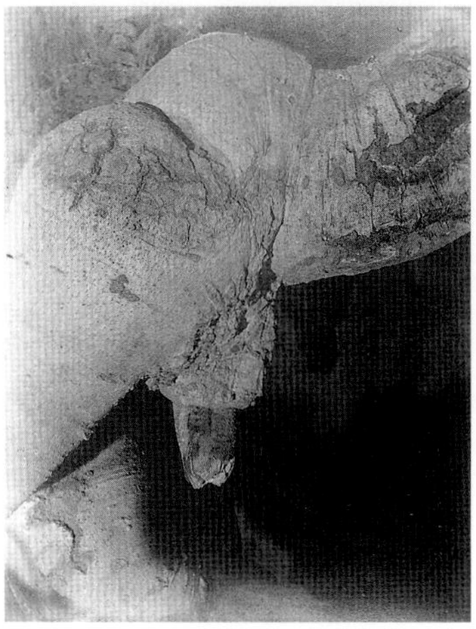

36. Imogen Cunningham: *Tree in the Sierra 3*, 1932

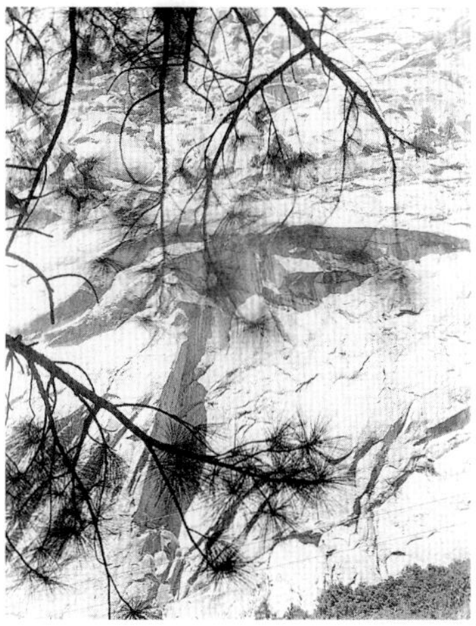

37. Imogen Cunningham: *Tree in the Sierra 4*, 1932

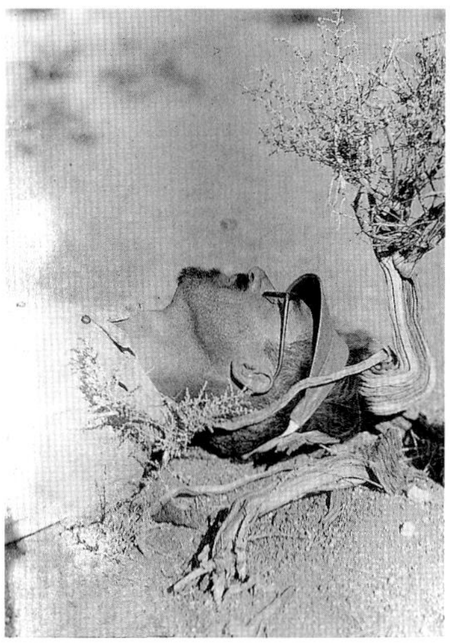

38. Imogen Cunningham: *Roi in the Alabamas*, 1923

the flower through three exposures (fig. 33). A fascinating example of her technique is found in a series of tree studies made in the Sierra in 1932. In a sample of four images, her style ranges from a classic composition of dead tree limbs against a cloud-strewn dark sky (fig. 34), recalling Alfred Stieglitz's contemporaneous studies of dying trees around his Lake George, New York, home; to an unusually composed, deep-focus juxtaposition of smooth and textured tree trunks (fig. 35); to a rare, humorous find of a tree trunk resembling a nude, contorted male (fig. 36); and then a double exposure of a rock cliff superimposed with a pattern of pine branches (fig. 37). The grouping exemplifies Cunningham's unique blend of spontaneous visual decisions and serendipitous attainment.

Surrealist irrationality, chance, and the extension of life beneath the earth are occasional elements of Cunningham's photography. Her appreciation for the absurd moment, or the surrealist equivalent of "the convulsive" and "the marvelous," is quite clear as early as 1923 when she captured Roi, at rest on a camping trip to the Alabamas, seemingly sprouting a sinewy plant form from his head (fig. 38). Root systems fascinated her, and she frequently documented them (as well as other forms of growth) with her own system of time-lapse photography. Just as aggressive and challenging plant forms captivated her attention, so too did their alternate, dark side—their otherworldly component. The clawlike shoots of an aloe grasping a mound of dirt have an eerie, fascinating power (fig. 39). These images clearly display plant morphology and growth patterns yet maintain an additional dimension that connotes the power and magic of an emanating life force. Edward Weston's *Cabbage Fragment* (fig. 40), looking as if it were carved from marble, projects a similar vigor. André Breton used an image of comparable force, Brassaï's sprouting potato (fig. 41), to illustrate a surrealist text. Like a trophy head or perhaps a ritual mask, the photograph was applied by Breton to his theory of "*magique-circonstancielle*" whereby "a found object is perceived as a sign of the finder's desire."[48] Cunningham's odd root study (fig. 42) suggests her own desire to aestheticize the underappreciated, the disagreeable, and the ugly. Perhaps paralleling her own self-image, she favored the unattractive for its more interesting idiosyncrasies.

Likewise, she admired natural forms that took their shape by chance, such as peculiar driftwood forms reduced by time, tides, and erosion. During the 1930s she photographed for Gump's department store in San Francisco a large collection of fluviana, wood pieces worn by rivers, oceans, and the elements, mounted on pedestals like precious sculptures. Their shapes resembled rabbits, fish, birds, demons, and dancing women in classical flowing robes—all magically modeled by the quirks of nature. (The writer James Joyce had, in fact, published in *transition* magazine [1929] a collection of photographs of fluviana resembling animals, feet, and shoes, all found in

Austria's Salzach River.) Cunningham delighted in discovering bizarre, identifiable forms in the workings of nature, and occasionally one of her stark overviews such as *Driftwood, Beach in Oakland* (fig. 43), produces the disorienting confusion of a Tanguy landscape. Or, like the paintings of the Italian illusionist Arcimboldo, a few of Cunningham's images reveal their subjects' curious anthropomorphic qualities. *Coconut* (fig. 44) is a simple but sly depiction of an alien creature, half-flora, half-fauna. Likewise, *Carrot* (fig. 45) seems more an impish gnome than a product of a vegetable garden. Its incongruous formality suggests it might have been a joke with which to tease Edward Weston.

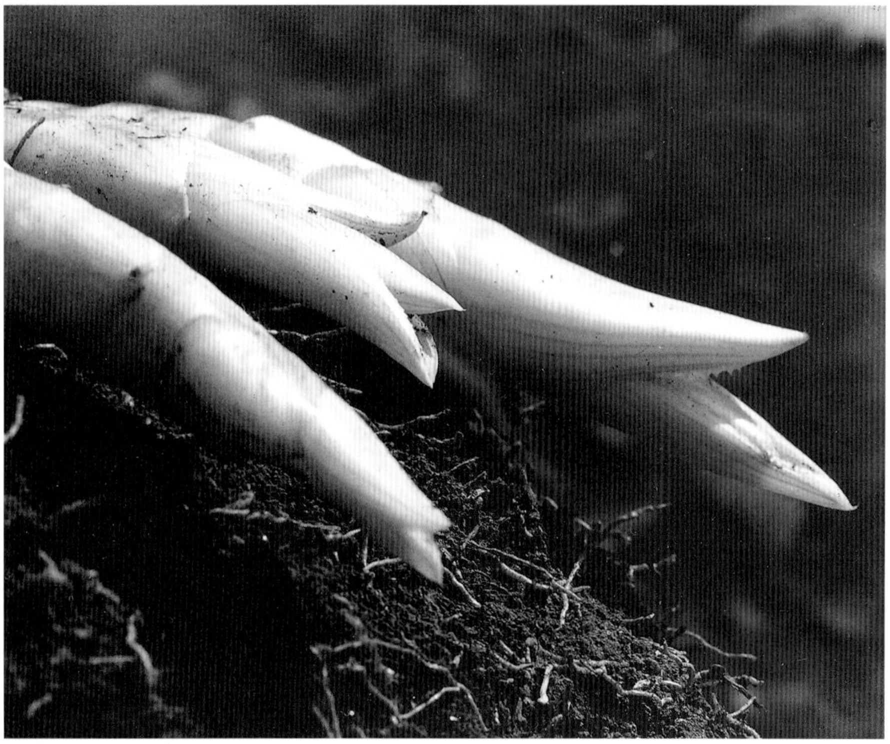

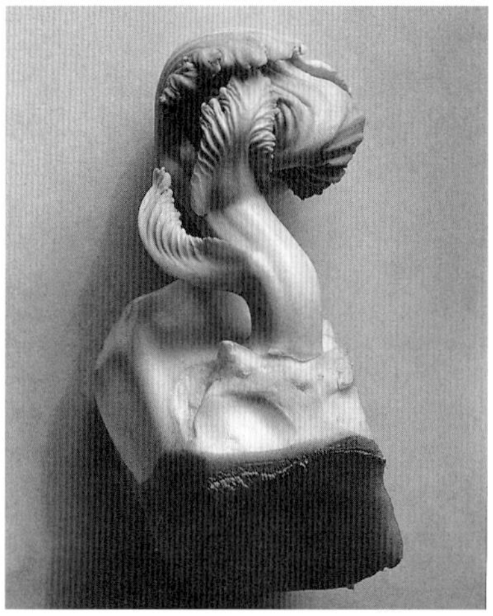

LEFT:
39. Imogen Cunningham: *Aloe Shoots,* 1920s

RIGHT:
40. Edward Weston: *Cabbage Fragment,* 1931. Copyright 1981 Center for Creative Photography, Arizona Board of Regents, Tucson.

Divorce from Roi Partridge in 1934 and the Great Depression created rough times for Cunningham. She was challenged to find new ways of marketing her work. Only photographers subsidized by government programs or regularly hired for commercial or advertising work enjoyed anything like a secure income. The art market for photography was far from proven; in fact, few museums collected photographs until decades later.

Cunningham attempted to explore interior design and public commissions, specifically the creation of large-scale photographic murals and screens. The trend toward mural decoration was undoubtedly stimulated by the achievements of the Mexican artist Diego Rivera as well as the plentitude of Works Progress Administration commissions executed by American

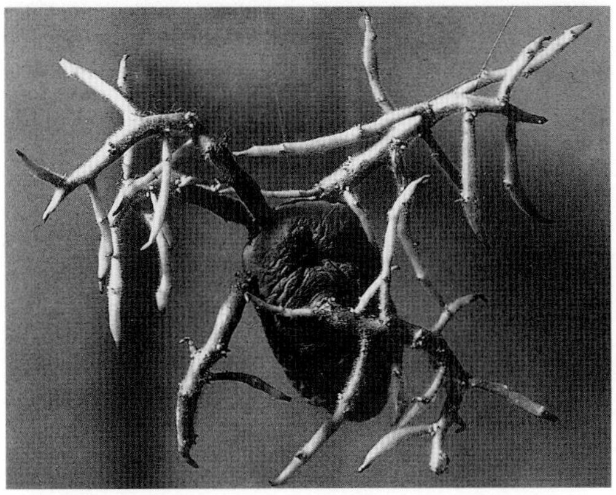

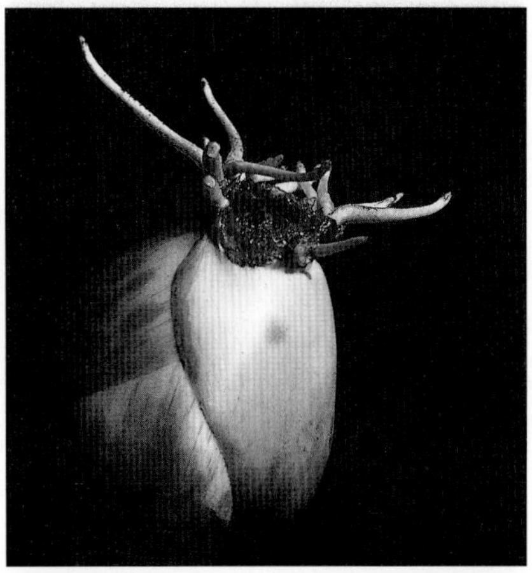

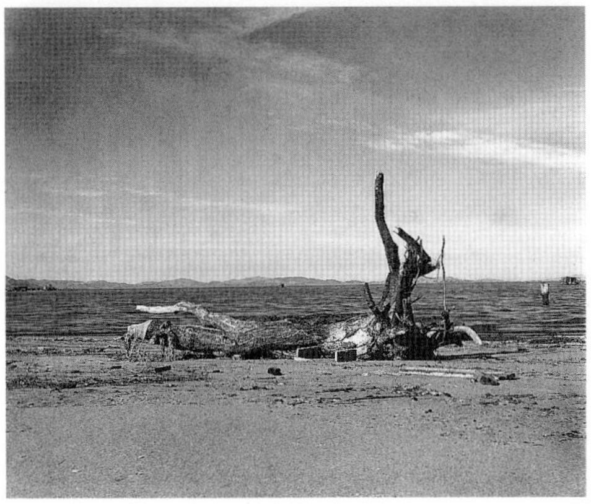

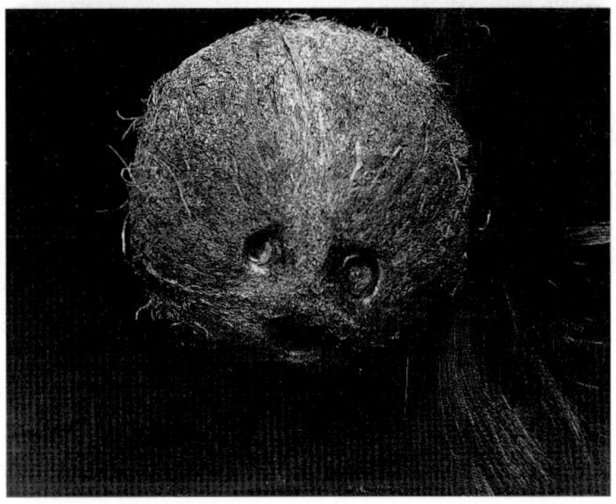

TOP LEFT, RIGHT:
41. Brassaï: *Magique-Circonstancielle,* 1931. Reproduced in *Minotaure,* no. 5 (1934).
42. Imogen Cunningham: *Root,* 1940s. Original in color (Kodachrome).

MIDDLE LEFT, RIGHT:
43. Imogen Cunningham: *Driftwood, Beach in Oakland,* 1930s
44. Imogen Cunningham: *Coconut,* 1920s

BOTTOM:
45. Imogen Cunningham: *Carrot,* 1940s. Original in color (Kodachrome).

artists in scores of public buildings. In 1932 the Museum of Modern Art, New York, mounted a major exhibition, "Murals by American Painters and Photographers," which included twelve photomurals, each 7 by 12 feet, by photographers such as Berenice Abbott, George Platt Lynes, Charles Sheeler, and Edward Steichen. Julien Levy, the New York gallery owner who specialized in surrealism and photography (and agent for Cunningham's work during the early 1930s), introduced the photographic works in the exhibition catalogue:

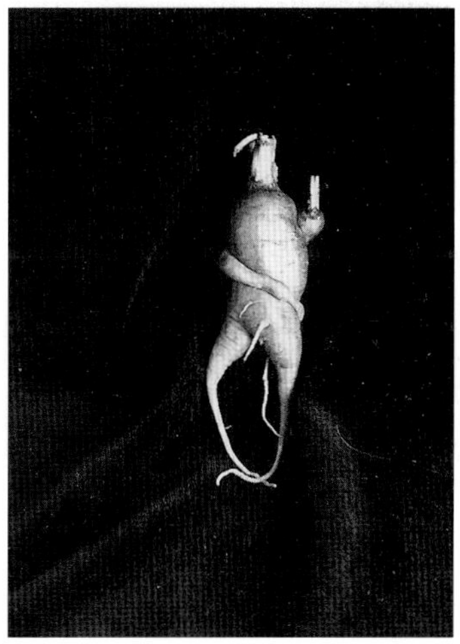

> The photographer is particularly well equipped to meet the problems of mural decoration as posed by the modern architect and builder. . . . A good photo-mural is not merely the mechanical enlargement of a small photograph. The enlarged mural is a new and independent production, and the photographer who does not visualize in advance the final scale of his picture will usually be surprised and dismayed by the results. . . . It is difficult to stretch a single, simple photograph over a large space and maintain interest, but it is dangerous to enlarge a complicated negative. . . . There exists no traditional solution to aid the photographer . . . [and] they must all answer one question in common: what can the photograph present that is not better rendered in paint?"[49]

Enlarged plant forms seemed a perfect subject for decorative uses, and Cunningham either previsualized new botanical images or adapted existing negatives to create numerous maquettes for murals and screens. She grouped four duplicates of an agave plant design (pl. 20) to form an energized large-scale pattern. In another technique, she created triptychs from two variant negatives of the same plant, such as *Flax* (pl. 7) and *Sedum Cristate* (pl. 46). The central image would be fuller and more complicated; the side panels were printed from the same negative, but with one side flipped, to produce a harmonious and balanced composition. Cunningham clearly labeled several images as designs for screens (pl. 88), although a lack of funds or commissioning clients seems to have prevented their production. *Colletia Cruciata, Design for a Screen* (fig. 46), Cunningham noted, could be "divided into 2 narrow and 1 wide panel."[50] A study of an intricate network of smoke tree branches *(Dalea spinosa)* became an evenly proportioned diptych, with one of its two identical halves rotated 180 degrees to complete the effect (fig. 47).[51]

Cunningham also turned to color film in her attempts to support herself. She solicited such magazines as *U.S. Camera, Life, Sunset, House and Garden,* and *Fortune* to include her color work. Her vivid, brilliantly colorful detail of fuchsia flowers graced the cover of *Sunset* magazine in June 1940 (pl. 78), but Cunningham's taste in plants sometimes proved too eccentric for general publications. When *Family Circle* requested a sampling of flower and garden pictures for an article, Cunningham eagerly suggested a selection of unique California species—"subtropical things which other states cannot manage"—including images of various succulents.[52] The art director of the magazine declined all her pictures, noting that "many of them were a little too unusual for the article we had in mind."[53]

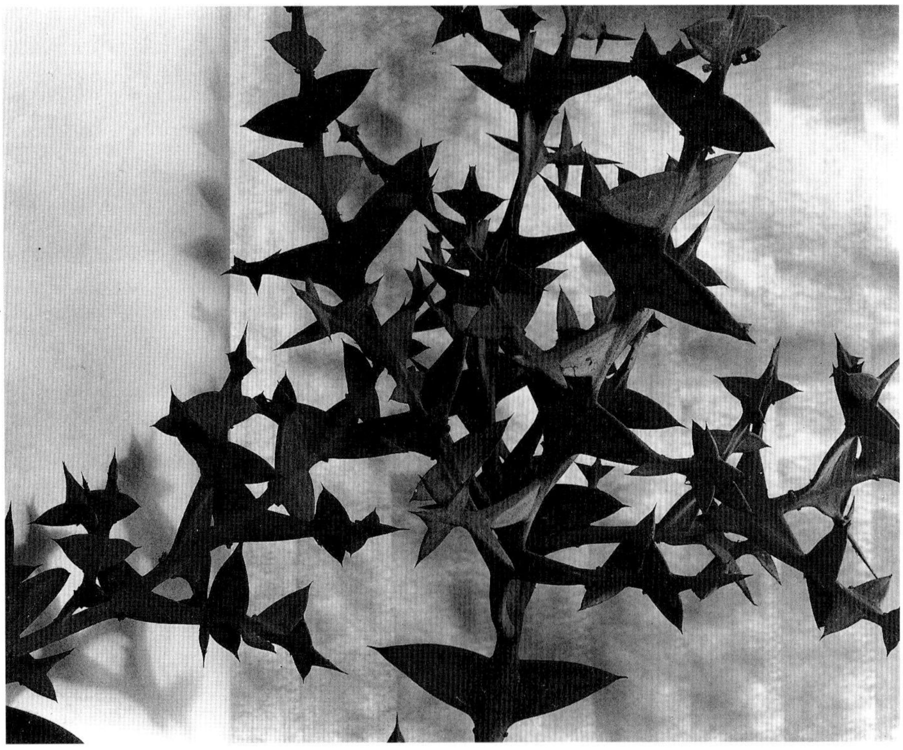

46. Imogen Cunningham: *Colletia Cruciata, Design for a Screen,* 1939

TOP:
47. Imogen Cunningham: *Smoke Tree*, ca. 1941. Unique diptych.

BOTTOM:
48. Imogen Cunningham: *1331 Green Street Garden*, 1951

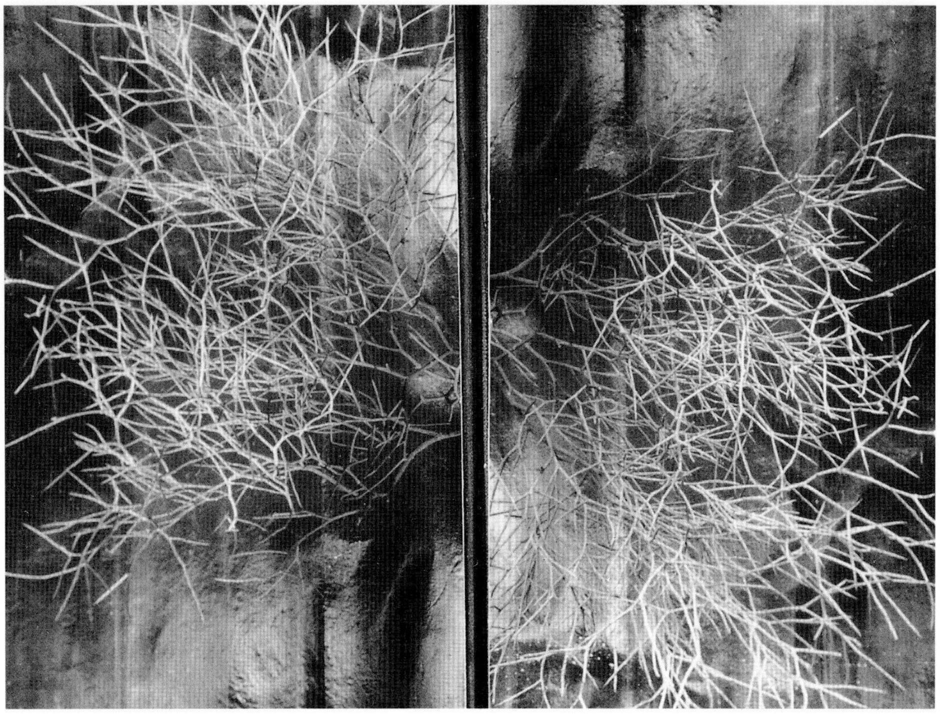

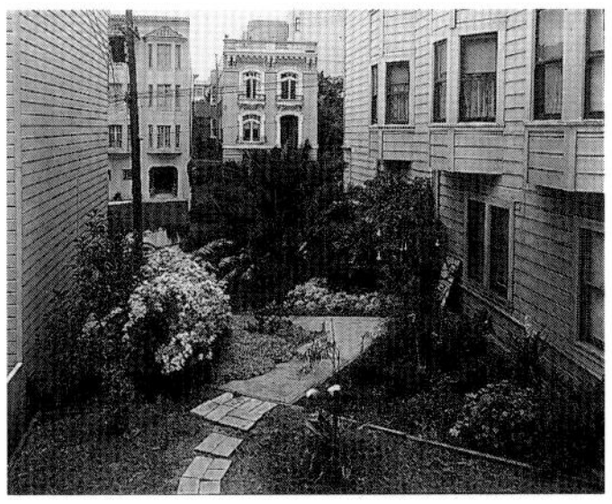

After World War II, Cunningham moved from Oakland to a small cottage in San Francisco. The tiny white house sat at the rear of a narrow lot abundantly planted with lush foliage and a palm tree (fig. 48). "I had the whole front yard designed by a landscaper [Eckbo, Royston and Williams] and paid in photography. That's one time when it was a good trade,"[54] she quipped. The house was only a few blocks from the California School of Fine Arts, where Cunningham taught in the photography department under the directorship of Minor White from 1947 to 1950.

Cunningham frequently used photography as an intuitive medium of transformation. On walkabouts through the city ghettos or treks through rural California landscapes, she extracted wonderful textures, patinas, and curiosities from rubble, neglect, and decay. Distinct from their inherent social commentary, her pictures became part autobiography, part travelogue, and part revelation. During an afternoon's wander through the barren terrain of a Sierra foothill, she captured her dark side—her shadow—encountering the stump of an ancient oak (fig. 49). The images are surreal with a hint of an "equivalent," Minor White's term for a photograph that transcends its original informational or pictorial purpose by evoking emotion.[55]

Perhaps because the courses she taught encompassed the history of photography and its technical developments, Cunningham after the late 1940s renewed her own experimental camera and darkroom techniques. She made a series of ghostly photograms of the exposed veined patterns of desiccated leaves by placing the leaves directly on the printing paper or by using them, like negatives, in her enlarger (fig. 50). This particular work strongly resembles

Man Ray's leaf photograms from the 1920s and those created by Max Ernst in collaboration with Man Ray in the 1930s (fig. 51). Ernst first made rubbings of leaf patterns (a technique he called *frottage*) to be used as negatives in contact printing. As in a photogram, the resulting reversal of tones emphasized the internal networks of the leaves, which André Breton compared to the nervous system.[56]

Cunningham also returned with enthusiasm to the double exposure. *Confusion in the Forest* (fig. 52) is a double-exposed negative of a eucalyptus trunk she encountered on a trip to Point Reyes. Central to the picture is the elliptical callus formed where a branch had been severed. Enhanced by its layering technique, the apparitional image connotes a dark entrance to a place of mystery. The sister technique of double exposure, printing with "sandwiched" negatives, became even more fascinating to Cunningham. It allowed her much more freedom to mix and match multiple negatives rather than previsualize an image onto one piece of film. Unlike photomontage, which involved rephotographing a cut and pasted photographic collage, exposing two or three layered negatives in the enlarger produced a finished print more efficiently. In a humorous study of hands intertwined with the fingerlike projections of her *Aloe plicatilis* plant (fig. 53), Cunningham combined two negatives, with one flipped, creating the illusion of a man-eating plant.

Cunningham never masked or manipulated her negatives in any labored way, and in this sense, she adopted a more dadaesque spirit of delight, shock, and mischievousness in the combinations and juxtapositions of her imagery. She paired a portrait of a ghostly looking man with a still life of fibrous bark (fig. 54) to compose what could well be a "spirit" photograph of a man

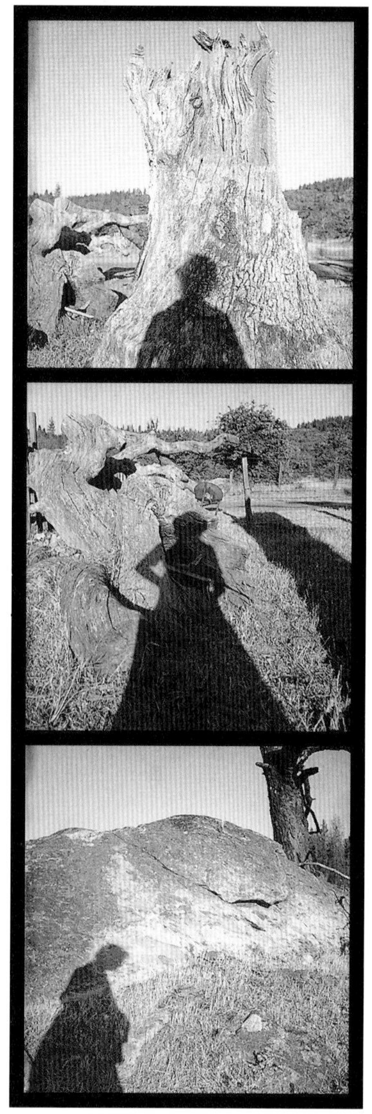

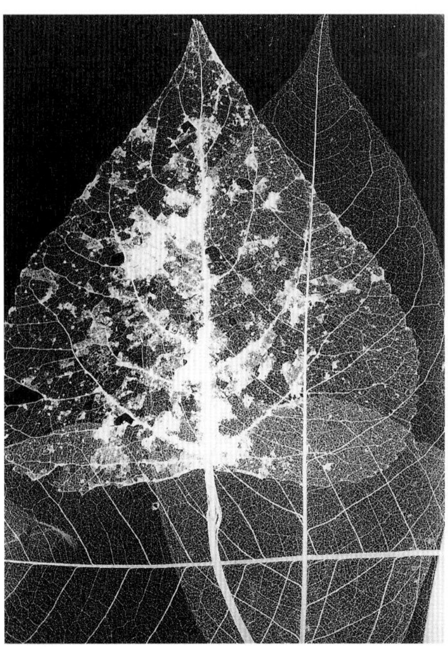

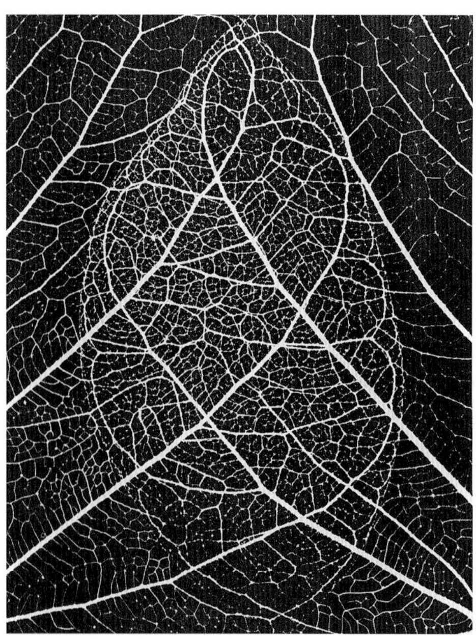

TOP:
49. Imogen Cunningham: *Self-portrait, Grass Valley,* 1946. Contact sheet.

LEFT:
50. Imogen Cunningham: *Leaves,* 1940s–1950s. Unique photogram enlargement.

RIGHT:
51. Max Ernst in collaboration with Man Ray: *. . . unir ton systeme nerveux au mien dans la nuit profunde de la conaissance,* 1936. Reproduced in *Minotaure,* no. 8 (1936).

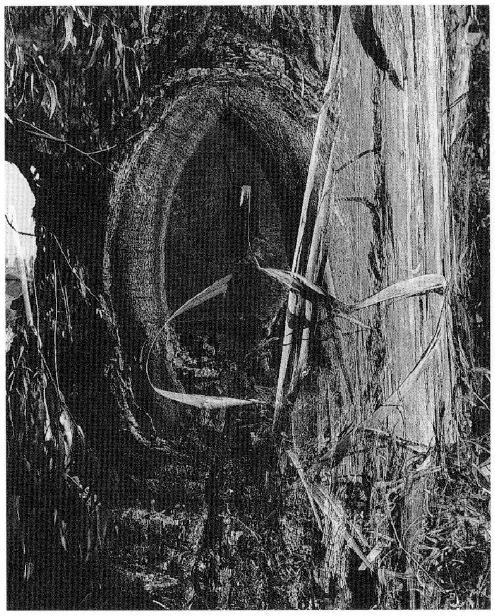

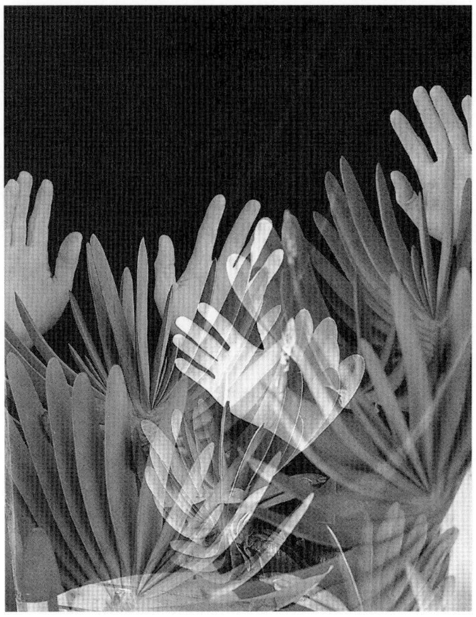

surrounded by ectoplasmic emissions. Cunningham often photographed genre textural images, such as craggy rock walls, fields of grasses, and finely branched tree limbs, which she kept as "doubles" negatives for future use. Portraits and body parts, often photographed against dark backgrounds, became the usual complementary image in the negative sandwich. Occasionally, some of Cunningham's images appear to be layered negatives when, in fact, they are directly photographed. An arrangement of desiccated leaves carefully placed on a glass table with her model below allowed Cunningham to create an evocative portrait study sprinkled with a shower of translucent leaves (fig. 55).

TOP LEFT, RIGHT:
52. Imogen Cunningham: *Confusion in the Forest,* 1951
53. Imogen Cunningham: *Hands with Aloe Plicatilis,* 1960

BOTTOM LEFT, RIGHT:
54. Imogen Cunningham: *Stuart Wheeler Double,* 1961
55. Imogen Cunningham: *Taiwan Leaves 2,* 1963

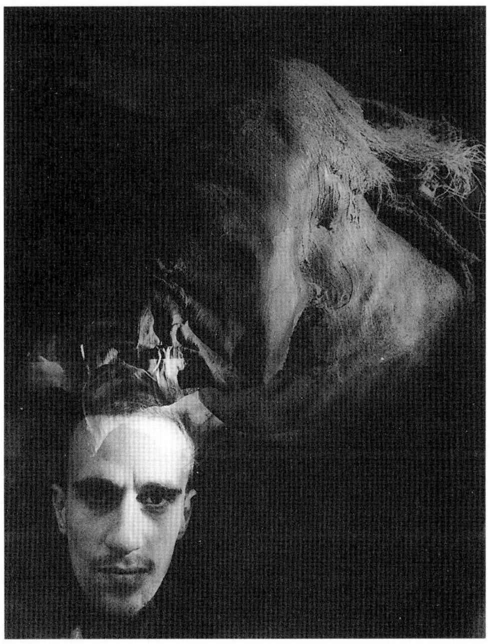

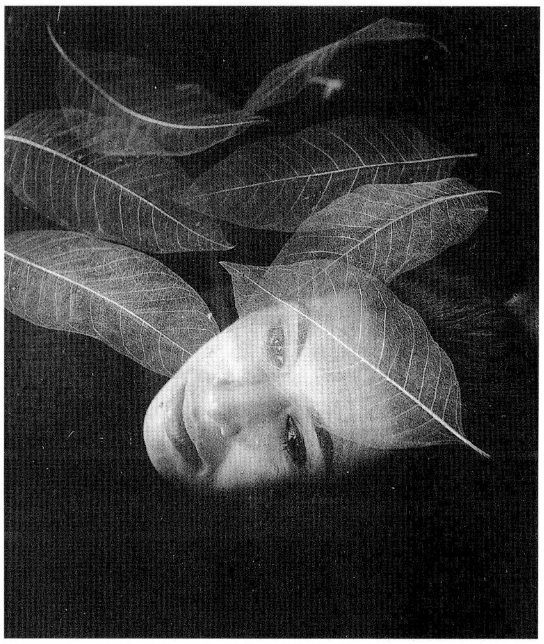

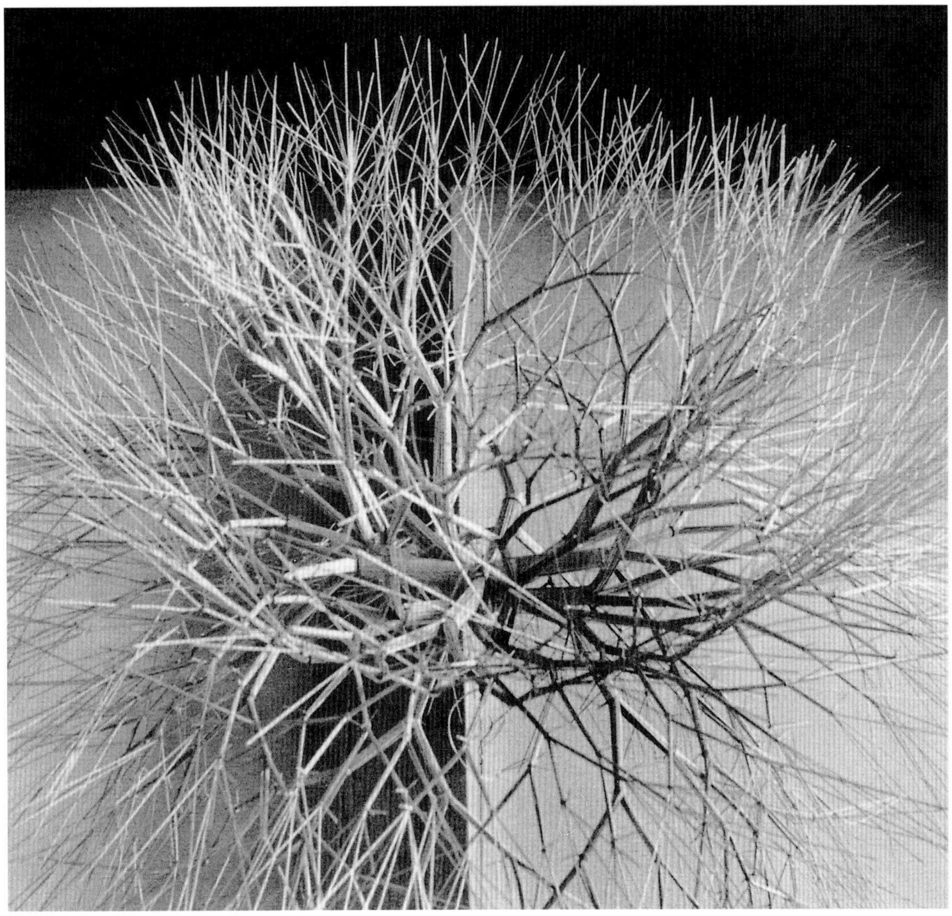

TOP:
56. Imogen Cunningham: *Ruth Asawa's New Expression with Metal,* 1963. Reproduced on the cover of *Aperture,* vol. 11, no. 4 (1964).

BOTTOM:
57. Imogen Cunningham: *Dream Walking,* 1968

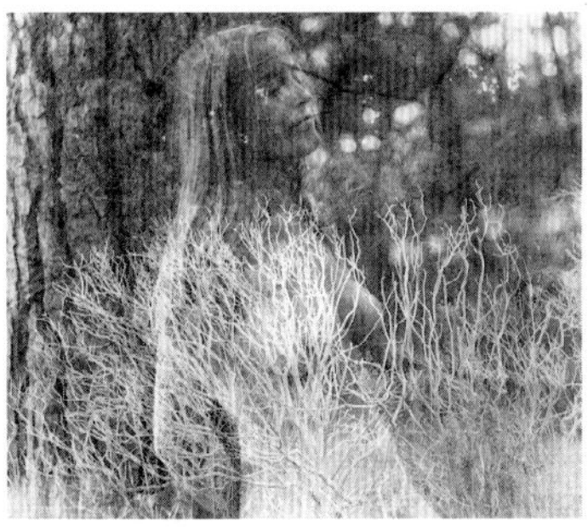

When the seminal photographic publication *Aperture* devoted its winter 1964 issue to Cunningham, editor Minor White aptly remarked in his preface: "Miss Cunningham shows the desirable characteristics of the hardy perennial."[57] On the front cover, White reproduced a quasibotanical image —a study Cunningham made of a shrublike, wire sculpture by artist Ruth Asawa (fig. 56), an astounding feat of translating line into three-dimensional form. Of the image's significance, White noted: "[It] set the pace for the rest—a full circle of life, experience and still whirling."[58]

The infinitely branching tree form was a powerfully expressive symbol for Cunningham, and she frequently incorporated the motif as an overlay in portraiture, as in *Dream Walking* (fig. 57), in which a nude model inhabits a symbolic forest landscape. "It seems to me," she wrote, "that each person finds his own symbols and also interprets them after his own fashion. Find them, create them, and interpret them. It is the other person who is going to come into difficulty."[59]

For the *Aperture* monograph, Cunningham eagerly tested the newly developed Polaroid film, provided in generous quantity by the Polaroid Corporation to create an image for the issue in honor of the company's financial support of the publication. She noted: "Quite a number of people have seen my tries and quite a few people like them. . . . Anyway it is a challenge and I would like to work more on this."[60] Throughout the 1960s, she occasionally used black-and-white and color Polaroid film for portraiture and still lifes (fig. 58). Imogen, ever experimental, accepted and utilized serendipitous solarization during chemical development in the 1960s, producing flashes of positive/negative reversal and variable sepia toning in her black-and-white prints. In a rare, unique photograph whose base image is a high-contrast silhouette of the crown of a pollarded San Francisco tree, she printed a double-exposed negative and then, during its development, solarized a random splash pattern through its branches (fig. 59).

In the late years of her life, Cunningham incorporated botanical imagery in her work, but previous ideas of regeneration often seem lost in a profound pessimism and focus on the ironies of old age. She photographed a handsome old woman with her philodendron plant trailing beside her, its stems bare and leggy and bearing only a few leaves, and called the picture *Age and Its Symbols* (1958). In a 1972 self-portrait, Imogen photographed her aged

LEFT:
58. Imogen Cunningham: *Lily*, 1960s. Unique Polaroid photograph.

RIGHT:
59. Imogen Cunningham: *Tree Near Temple Emmanuel*, 1965. Unique solarized gelatin silver print.

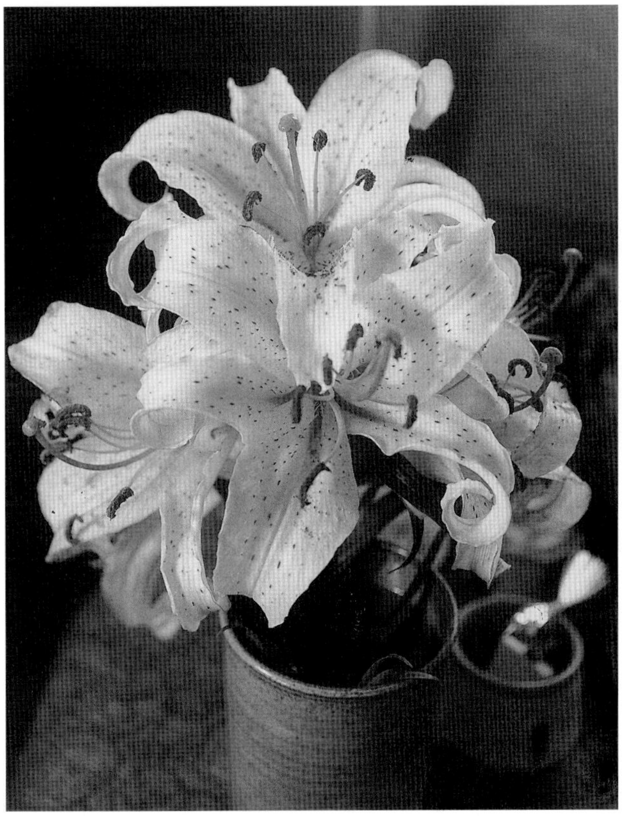

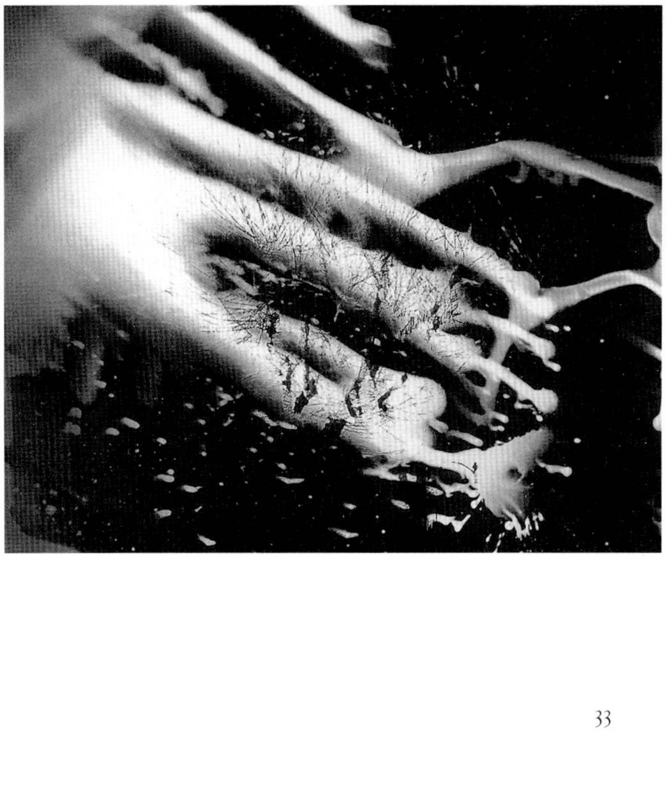

60. Imogen Cunningham: *Pentimento*, 1973

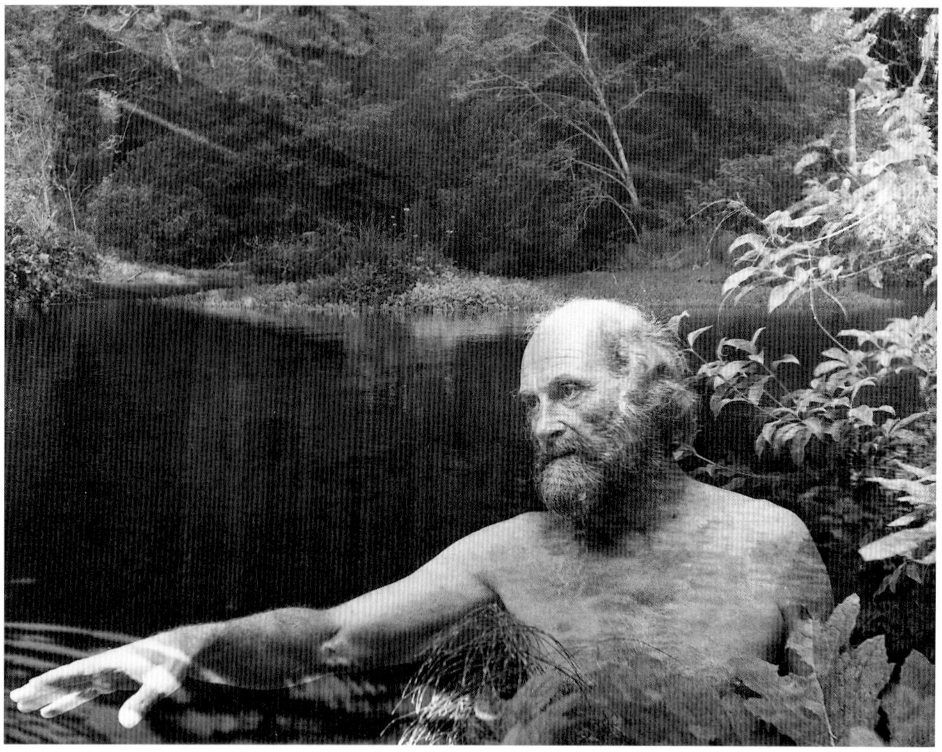

hand held high over the oddly shaped leaf of a voodoo lily plant (pl. 100). Comparing her wrinkled hand to the structure of the leaf, however, is only part of her point. One needs to know that the scent of the bloom of this unusual specimen is repellent, like rotten flesh.

One of Cunningham's last great and meaningful nature photographs is a portrait of the artist Morris Graves. In 1973 Imogen spent a day at his compound in Loleta, California. Influenced by the introduction to Lillian Hellman's just published literary portraits, *Pentimento* (1973), Imogen fused separate images of the artist and his forested landscape in an evocative composition (fig. 60). Graves emerges from the trees like a conjurer of natural forces, or even perhaps as the life-giving figure of God on Michelangelo's Sistine Chapel ceiling. Graves is psychically incorporated into the landscape —a symbiotic fusion—and Hellman's description of pentimento as "a way of seeing and then seeing again"[61] applies perhaps to Cunningham's own insight into the importance of the relationship of man and nature. Possibly speculating in her imagery on her own significance within the universe, Cunningham, by the time of her death in 1976, might have crossed a new threshold of enlightenment.

Cunningham's botanical images are a testament to her curiosity, love of nature, and supreme photographic talents. Acknowledged as important icons of the history of twentieth-century photography, these pictures will remain timeless. In their diversity she has captured the uncanny laws of nature, so well described by Maeterlinck:

> This vegetable world, which to us appears so placid, so resigned, in which all seems acquiescence, silence, obedience, meditation, is, on the contrary, that in which impatience, the revolt against destiny are the most vehement and stubborn. . . . [A plant] exerts itself wholly with one sole aim: to escape above from the fatality below, to evade, to transgress the heavy and somber law, to set itself free, to shatter the narrow sphere, to invent or invoke wings, to escape as far as it can, to conquer the space in which destiny encloses it, to approach another kingdom, to penetrate into a moving and active world. . . . We shall see that the flower sets man a prodigious example of insubmission, courage, perseverance, and ingenuity. If we had applied to the removal of various necessities that crush us, such as pain, old age, and death, one-half of the energy displayed by any little flower in our gardens, we may well believe that our lot would be very different from what it is.[62]

Imogen Cunningham's flora eulogize her well and pay tribute to her into eternity. Remembering a Noguchi haiku, her scattered ashes murmur:

> Where the flowers sleep,
> Thank God I shall sleep, to-night.
> Oh, come, butterfly![63]

NOTES

The following short forms are used throughout the notes:

Cunningham/Daniel = Imogen Cunningham, interview by Edna Tartaul Daniel, June 1959, transcript, Regional Oral History Project, University of California Berkeley

ICA = Imogen Cunningham Archives, The Imogen Cunningham Trust, Berkeley, California

ICP = Imogen Cunningham Papers, Archives of American Art, Smithsonian Institution

1. William Blake, "VIII. Auguries of Innocence," from *The 'Pickering Manuscript,'* in *The Poems of William Blake,* ed. W. H. Stevenson, text by David V. Erdman (London: Longman Group; New York: W. W. Norton, 1971), p. 585, lines 1-4.

2. Ralph Waldo Emerson, "Nature," in *The Conduct of Life, Nature & Other Essays* (London: J. M. Dent and Sons; New York: E. P. Dutton, 1908), p. 3.

3. Imogen Cunningham, undated lecture notes, ICA.

4. E. Isabelle Halderman, "Successful Seattle Business Women: Miss Imogene [*sic*] Cunningham, *Seattle Post-Intelligencer* (1913), clipping, ICA.

5. Cunningham/Daniel, p. 51.

6. Cunningham/Daniel, pp. 84a–85.

7. Flora Huntley Maschmedt, "Imogen Cunningham—An Appreciation," *Wilson's Photographic Magazine* 51, no. 3 (March 1914), pp. 97–98.

8. Maurice Maeterlinck, "Field Flowers," reprinted in *The Double Garden,* trans. Alexander Teixeira de Mattos (New York: Dodd, Mead, 1915), p. 229.

9. Yone Noguchi, "The Garden of Truth" from *Seen and Unseen* (1897), in *Selected Poems of Yone Noguchi: Selected by Himself* (Boston: The Four Seas Company; London: Elkin Mathews, 1921), p. 8.

10. Roi Partridge changed the spelling of his name from "Roy" in 1919.

11. P. H. Emerson, quoted in *Life and Landscape: P. H. Emerson, Art and Photography in East Anglia 1885–1900,* ed. Neil McWilliam and Veronica Sekules (Norwich, England: Sainsbury Centre for Visual Arts, University of East Anglia, 1986), p. 21.

12. Imogen Cunningham, "One Hand in the Dishpan, The Other in the Darkroom," an undated lecture(?) note (internal evidence suggests an early 1940s date), ICA. Cunningham compared her approach to career and motherhood with that of Dorothea Lange: "I know of another woman who solved the same problem in an even more practical way. She too was a photographer. She had two children and always made enough money with her photography to replace herself in the household. She is now one of the best known documentary workers in the country."

13. Cunningham/Daniel, pp. 26–27.

14. The floral paintings of Georgia O'Keeffe are often but inappropriately cited as influencing Cunningham in the mid to late 1920s, but Cunningham would not see them until 1934. An article by Paul Rosenfeld, "The Paintings of Georgia O'Keeffe," had appeared in *Vanity Fair* 19, no. 2 (Oct. 1922), p. 56, but reproduced only two abstracted landscapes and a still life of apple forms. When Cunningham photographed Alfred Stieglitz at his gallery, An American Place, in 1934, she placed him next to Georgia O'Keeffe's masterpiece, *Black Iris* (1926), an image that closely resembles her own study of an iris from the 1920s. About the time of Cunningham's visit, Stieglitz had mounted an exhibition of twenty-eight of O'Keeffe's paintings from 1919 to 1934.

15. Cunningham preferred the more closely observed *Tower of Jewels.* She stated in a letter to Minor White, then negotiating to buy a collection of her work for the George Eastman House, that regarding "the one of the center of the same flower [magnolia] done at the same time . . . I like it better but maybe it isn't better" (Cunningham to Minor White, March 3, 1957, ICA).

16. See "Johan Hagemeyer," *The Archive* 16 (June 1982), monograph in the Research Series of the Center for Creative Photography, University of Arizona, Tucson, with essays by Richard Lorenz, John P. Schaefer, and Terence R. Pitts.

17. Jack Napton, "Incredible, Unforgettable James West, Plantsman (1886–1939)," *California Horticultural Society Journal* 28, no. 3 (July 1967), pp. 180–86. West's life had an unfortunate end. In 1936 he accompanied Dr. T. Harper Goodspeed, director of the University of California Botanical Garden, on a specimen collecting trip to Peru. He mysteriously disappeared but was seen in Berlin in 1938. His body was found in 1939 in the River Spree; he was possibly the victim of Nazi thugs who thought him to be an American spy; see ibid., pp. 187, 196.

18. Ibid., p. 181.

19. James West, "Silverskin," *Journal of the Cactus and Succulent Society of America* (Los Angeles) 1, no. 3 (Sept. 1929), p. 3.

20. James West, "Two Asclepiads," *Journal of the Cactus and Succulent Society of America* (Los Angeles) 1, no. 5 (Nov. 1929), p. 2.

21. Cunningham to Jo Sinel, Sept. 30, 1964, ICA.

22. Amédée Ozenfant, *Foundations of Modern Art,* trans. John Rodker (New York: Dover, 1952), pp. 57–58.

23. Carl Georg Heise, "The World Is Beautiful," in *Albert Renger-Patzsch, 100 Photographs* (Paris: Créatis; Cologne: Schürmann and Kicken, 1979), p. 10.

24. László Moholy-Nagy, "A New Instrument of Vision," in "From the Bauhaus," *Camera* (Switzerland) 46, no. 4 (April 1967), p. 30.

25. Karl Blossfeldt, *Art Forms in Nature, Second Series: Examples from the Plant World Photographed Direct from Nature by Professor Karl Blossfeldt* (New York: E. Weyhe, 1932), foreword (n.p.).

26. Walter Benjamin, quoted in Rolf Sachsse, *Karl Blossfeldt Photographs* (Cologne: Benedikt Taschen, 1994), p. 42.

27. Imogen Cunningham, undated note, ICA.

28. Cunningham to Alfred Frankenstein, Feb. 26, 1965, ICA.

29. Florence Lehre, "Artists and Their Work," *Oakland Tribune,* Oct. 27, 1929, p. S-7. The other photographers exhibiting were Brett Weston, Manuel Alvarez Bravo, Tina Modotti, Dorothea Lange, Roger Sturtevant, Anton Bruehl, E. A. Nievera, Ira Martin, and three members of the Japan Camera Club of Los Angeles, T. K. Shindo, R. Itano, and K. Nakamura.

30. Edward Weston to Cunningham, undated (late 1920s), ICA.

31. Edward Weston to Cunningham, Jan. 12, 1928, ICA.

32. Edward Weston, "Imogen Cunningham, Photographer," *The Carmelite,* April 17, 1930, p. 7.

33. Cunningham's photographs, titled in German, appeared in the original exhibition checklist as *Akt; Wassertank; Formen einer Blume* (pl. 11); *Pflanzenform* (pl. 27); *Schwarze und weisse Lilien* (pl. 42); *Calla* (pl. 8); *Pflanzenform* (pl. 24); *Pflanzenform* (pl. 37); *Eislilie* (pl. 22); and *2 Callas* (pl. 10). Cunningham occasionally used these German titles through the years. For *Film und Foto,* see *International Ausstellung des Deutschen Werkbundes, Film und Foto, Stuttgart 1929,* facsimile exh. cat. (Stuttgart: Deutsche Verlags-Anstalt, 1979).

34. Cunningham to Mary Sands, Jan. 7, 1931, ICA.

35. Junius Cravens, "The Art World," *The Argonaut,* Dec. 25, 1931, p. 8. A slightly edited version of the same exhibition, billed as *The Seeing Eye,* was

shown in the spring of 1932 at the Los Angeles Museum where it was noted: "This exhibit points out in most unmistakable terms how the speeding of modern life seems to have robbed us of the fun of seeing, —but that with such opportunities as this everyone may again develop an intimate contact with the untold beauties to be found in unexpected aspects of nature"; see "'Impressions in Silver' by Imogene [*sic*] Cunningham," *Los Angeles Museum Art News* (April 1932).

36. Ansel E. Adams, "Photography," *The Fortnightly* 1, no. 12 (Feb. 12, 1932), p. 26.

37. Imogen Cunningham, quoted in Paul Hill and Thomas Cooper, *Dialogue with Photography* (New York: Farrar, Straus, Giroux, 1979), p. 306.

38. Man Ray, "The Age of Light," in *Photographs by Man Ray, 1920 Paris 1934* (Hartford, Conn.: James Thrall Soby, 1934), n.p.

39. Edward Weston, *The Daybooks of Edward Weston,* vol. 2, *California,* ed. Nancy Newhall (Millerton, N.Y.: Aperture, 1973), p. 181.

40. Ibid., p. 203.

41. Ibid.

42. The original statement was included in *Das Deutsche Lichtbild* (1927) and quoted in Beaumont Newhall, "Albert Renger-Patzsch," *Image, the Journal of the George Eastman House of Photography* 3 (Sept. 1959), p. 142.

43. Heise, "The World Is Beautiful," p. 13.

44. Norman Bel Geddes, *Horizons* (Boston: Little, Brown, 1932), pp. 12–13.

45. For a more complete discussion of the organization and its influence, see the exhibition catalogue for the revival of the original show: *Group f.64* (St. Louis: University of Missouri–St. Louis, 1978), and Therese Thau Heyman, ed., *Seeing Straight: The f.64 Revolution in Photography* (Oakland, Calif.: The Oakland Museum, 1992).

46. Franz Roh, "Mechanism and Expression, The Essence and Value of Photography," in *photo-eye,* ed. Franz Roh and Jan Tschichold (Stuttgart: Akademischer Verlag Dr. Fritz Wedekind, 1929), p. 16.

47. Ibid.

48. Dawn Ades, "Photography and the Surrealist Text," in Rosalind Krauss and Jane Livingston, *L'Amour Fou: Photography and Surrealism* (Washington, D.C.: Corcoran Gallery of Art; New York: Abbeville Press, 1985), p. 183.

49. Julien Levy, "Photo-murals," in *Murals by American Painters and Photographers,* exh. cat. (New York: Museum of Modern Art, 1932), pp. 11–12.

50. Imogen Cunningham, notations on verso of vintage 8″ × 10″ photograph, ICA.

51. While vacationing in Palm Springs in 1941, Cunningham had sent László Moholy-Nagy some samples of this deciduous desert tree. He responded on School of Design in Chicago letterhead: "Heartiest thanks for your lovely greetings with a bit of smoke tree from Palm Springs. I can imagine how wonderful it is to have a little rest. . . . I always show your pictures in my lectures with great pleasure both to myself and to the audience and I carry in my memory your splendid thoughts, bon mots and other manifestations of your fine work" (Moholy-Nagy to Cunningham, Jan. 4, 1941, ICP, roll 1633).

52. Cunningham to Jan C. Mayer, Jan. 26, 1955, ICP, roll 1633.

53. Jan C. Mayer to Cunningham, Aug. 10, 1955, ICP, roll 1633.

54. Imogen Cunningham, in Hill and Cooper, *Dialogue with Photography,* p. 305.

55. See Minor White and Walter Chapell, "Some Methods for Experiencing Photographs," *Aperture* 5, no. 4 (1957), pp. 156–71.

56. Ernst's photogram and frottage illustrated André Breton's book *Le Château étoilé* (1936). The photogram also was used to illustrate Breton's article, "Le Château étoilé," in *Minotaure,* no. 8 (1936), p. 34.

57. Minor White, "Imogen Cunningham," *Aperture* 11, no. 4 (1964), p. 2. The issue also included a biographical essay on Cunningham by George Craven.

58. Minor White, quoted in Cunningham to George Craven, April 9, 1964, ICP, roll 1634.

59. Imogen Cunningham, lecture notes for University of California Extension, Oct. 27, 1968, ICA.

60. Cunningham to Ansel Adams, Feb. 10, 1964, Center for Creative Photography, University of Arizona, Tucson.

61. Lillian Hellman, *Pentimento: A Book of Portraits* (Boston: Little, Brown, 1973), p. 3.

62. Maurice Maeterlinck, "The Intelligence of the Flowers," reprinted in *The Measure of the Hours,* trans. Alexander Teixeira de Mattos (New York: Dodd, Mead, 1907), pp. 249–51.

63. Yone Noguchi, "X," from *Japanese Hokkus* (1920), in *Selected Poems of Yone Noguchi,* p. 125.

Plates

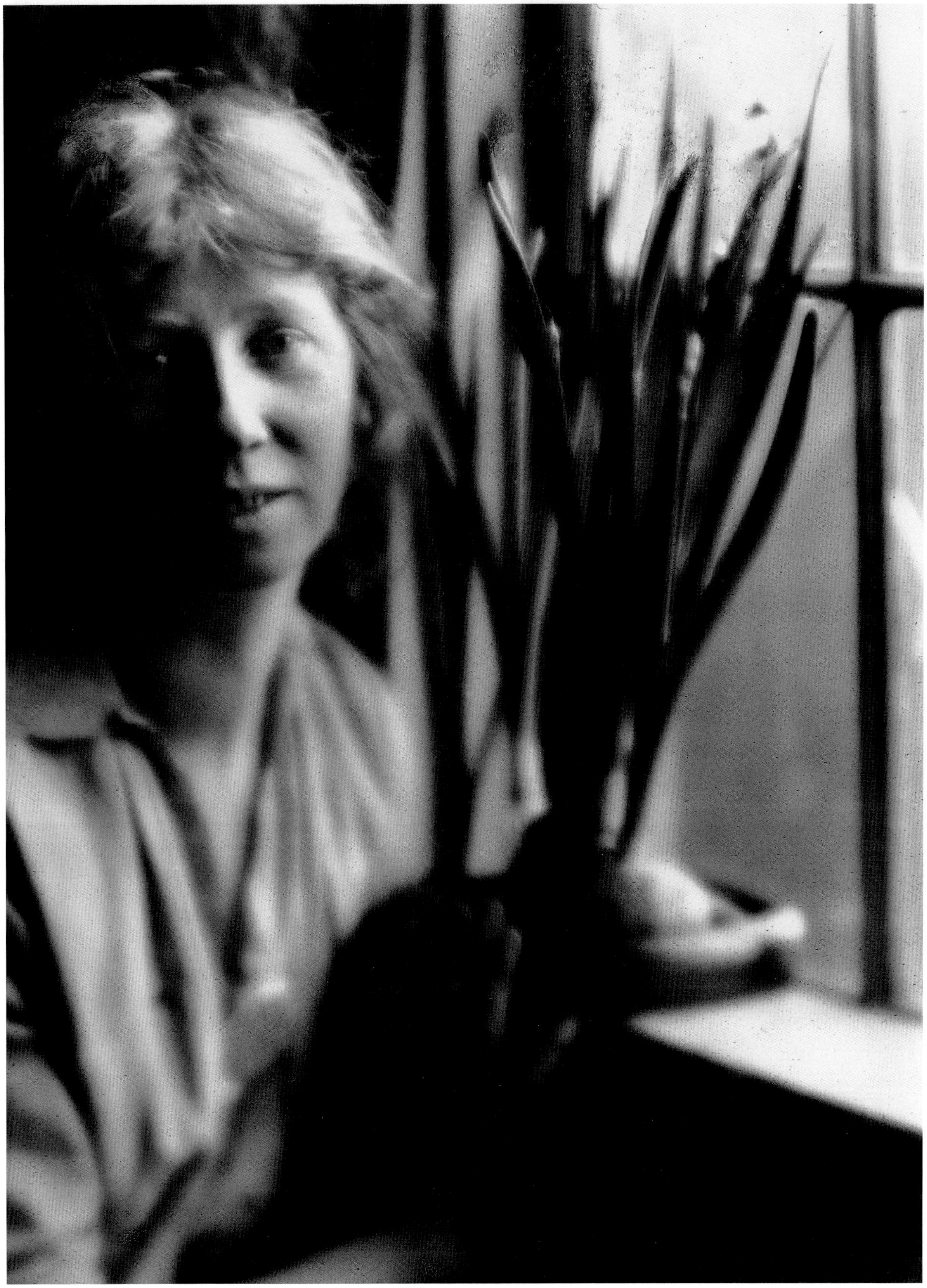

PLATE I

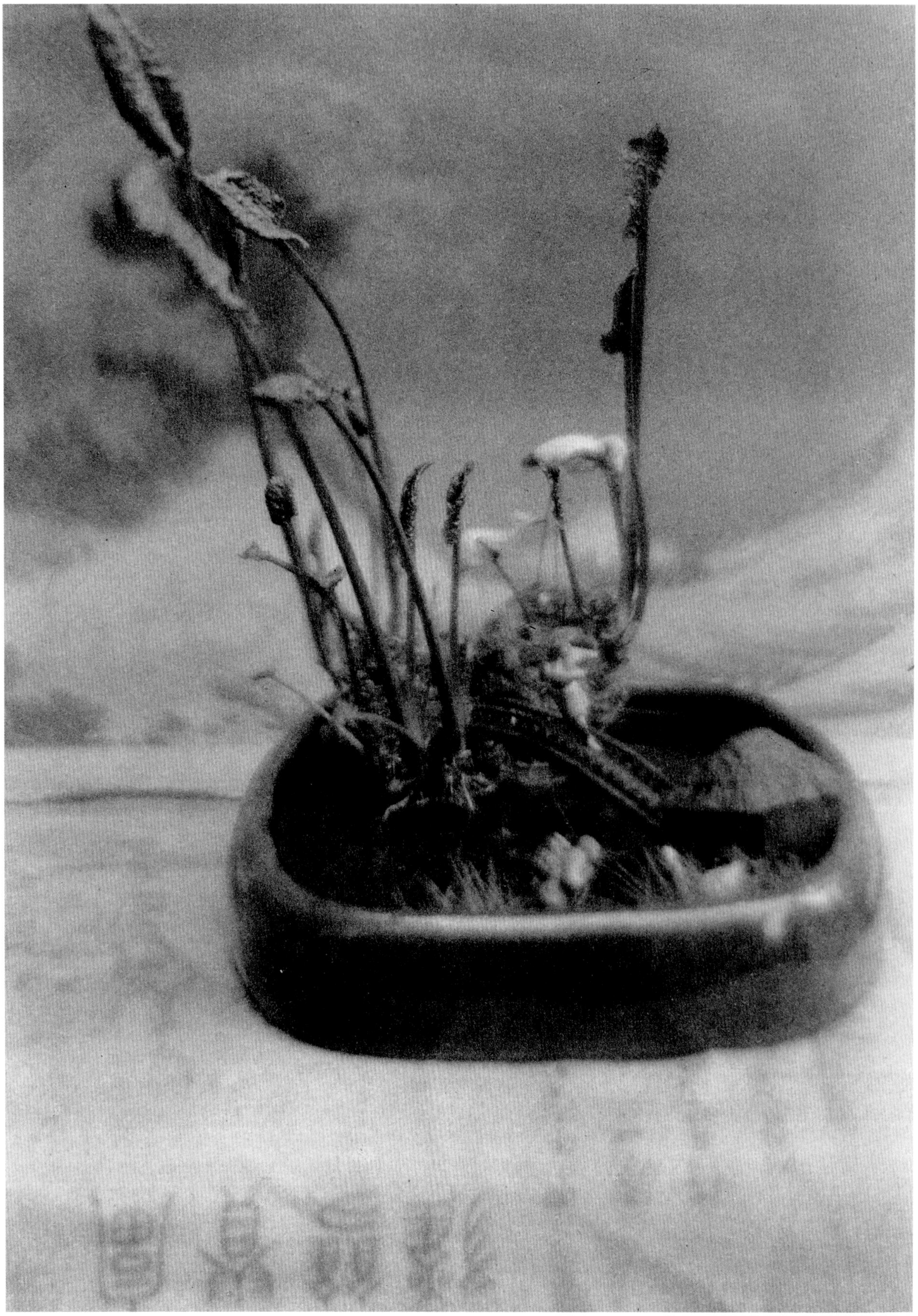

PLATE 2

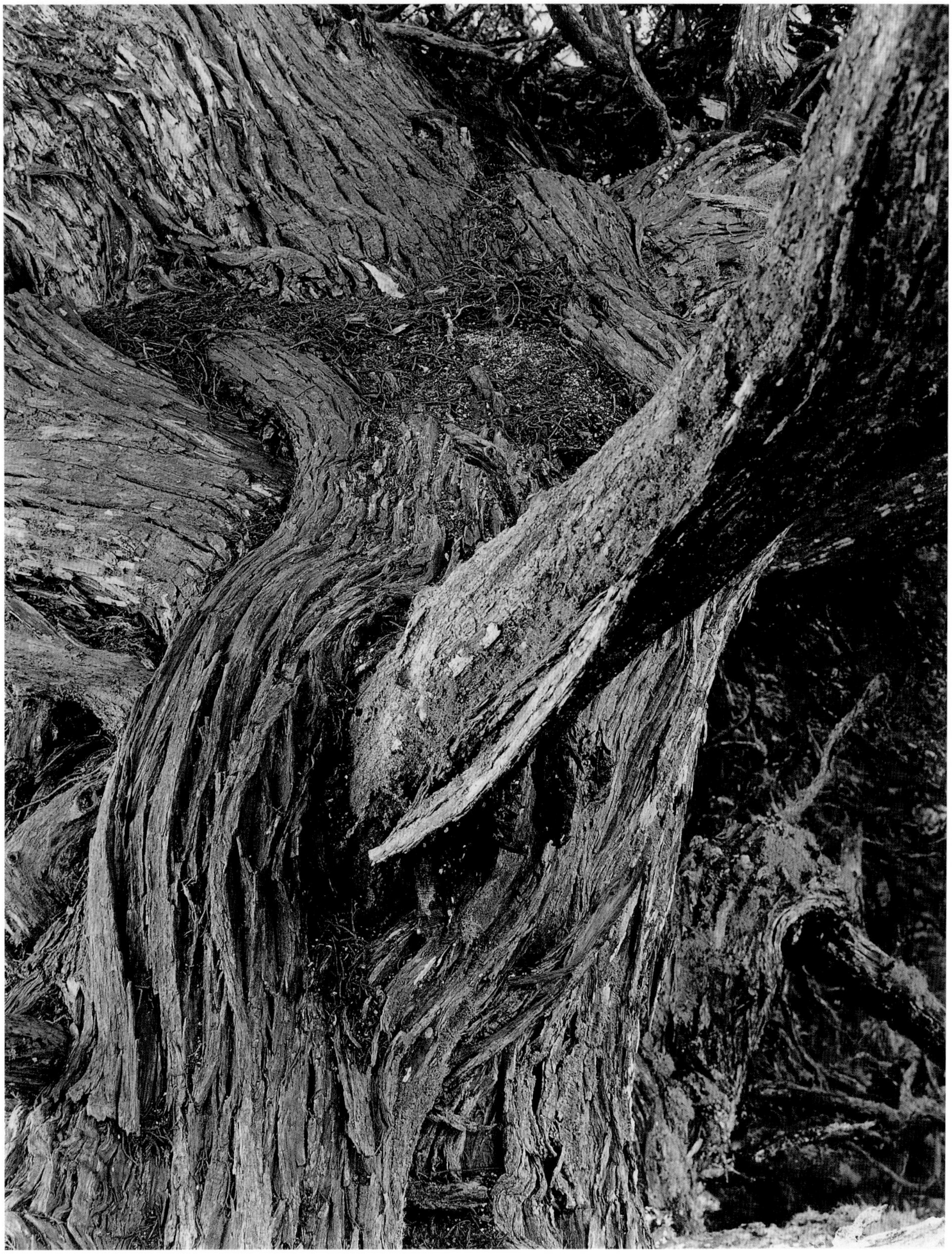

PLATE 3

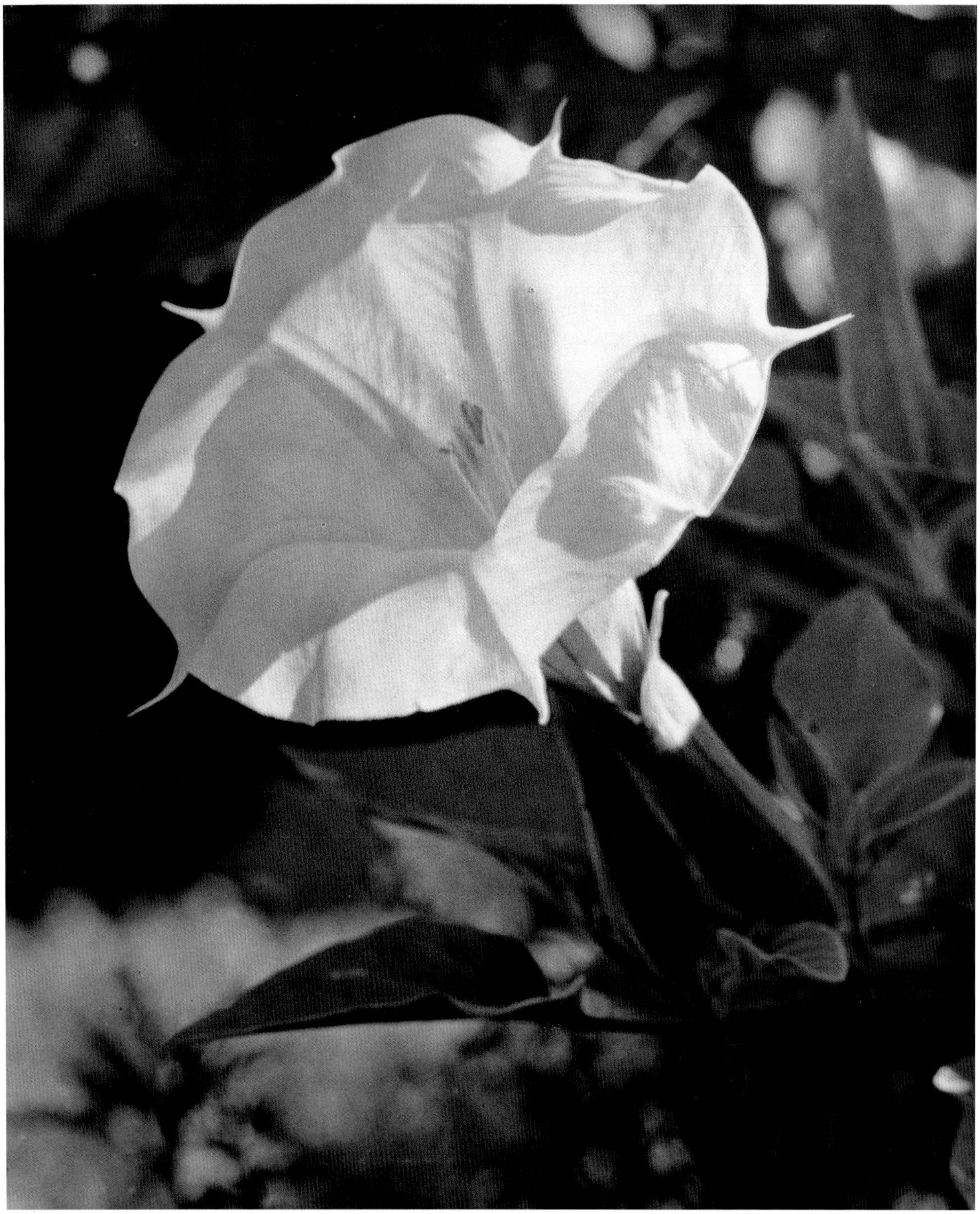

PLATE 4

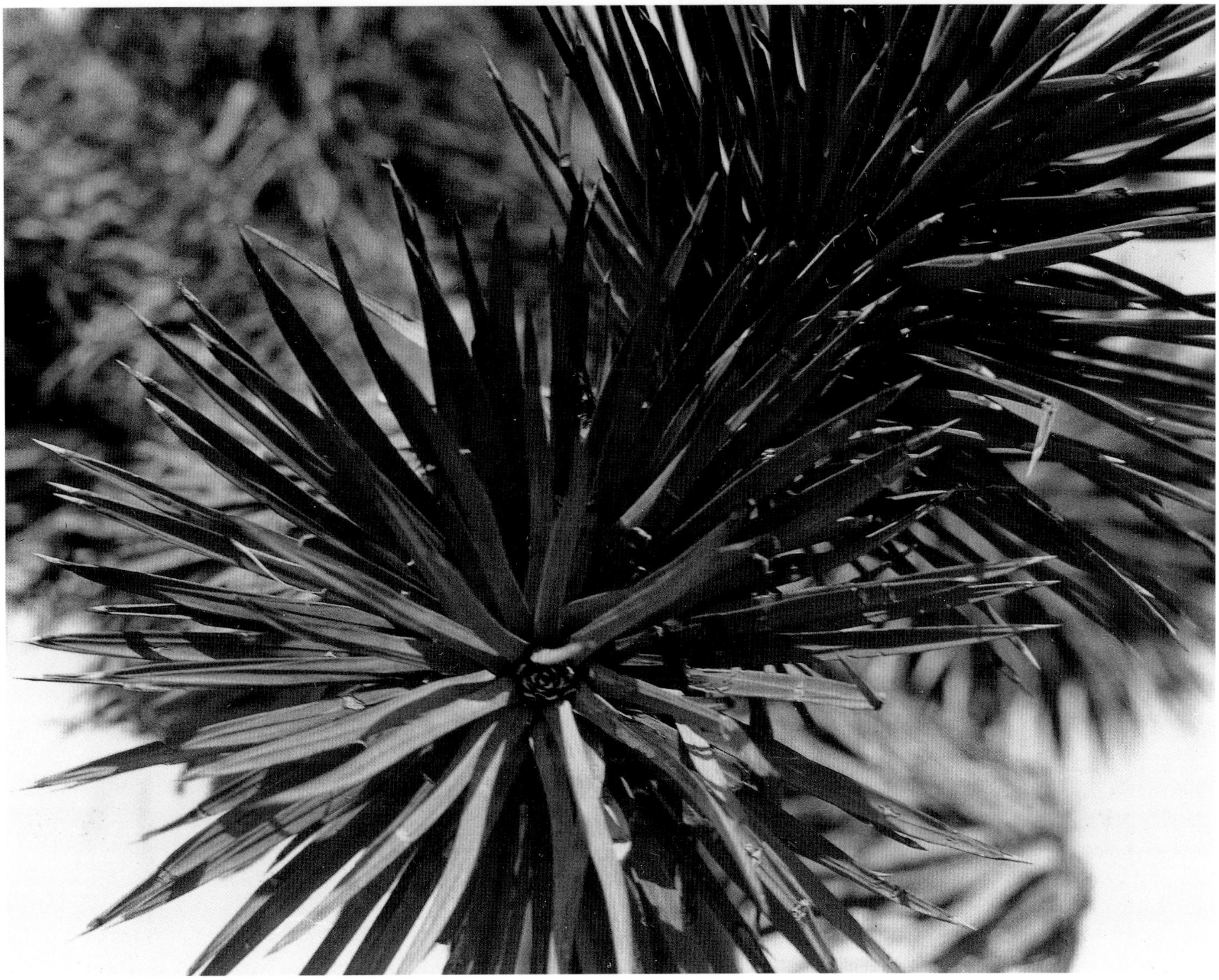

PLATE 5

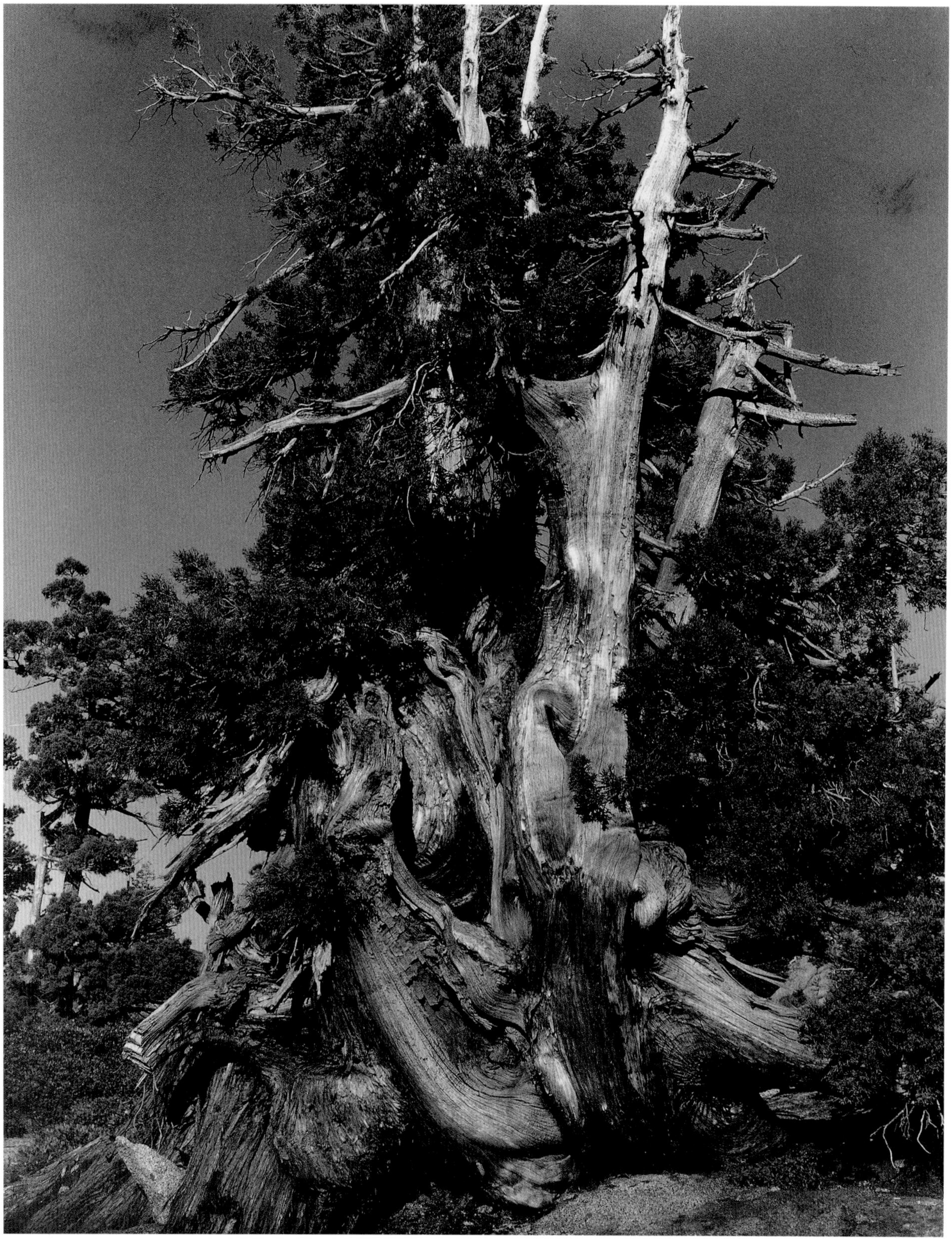

PLATE 6

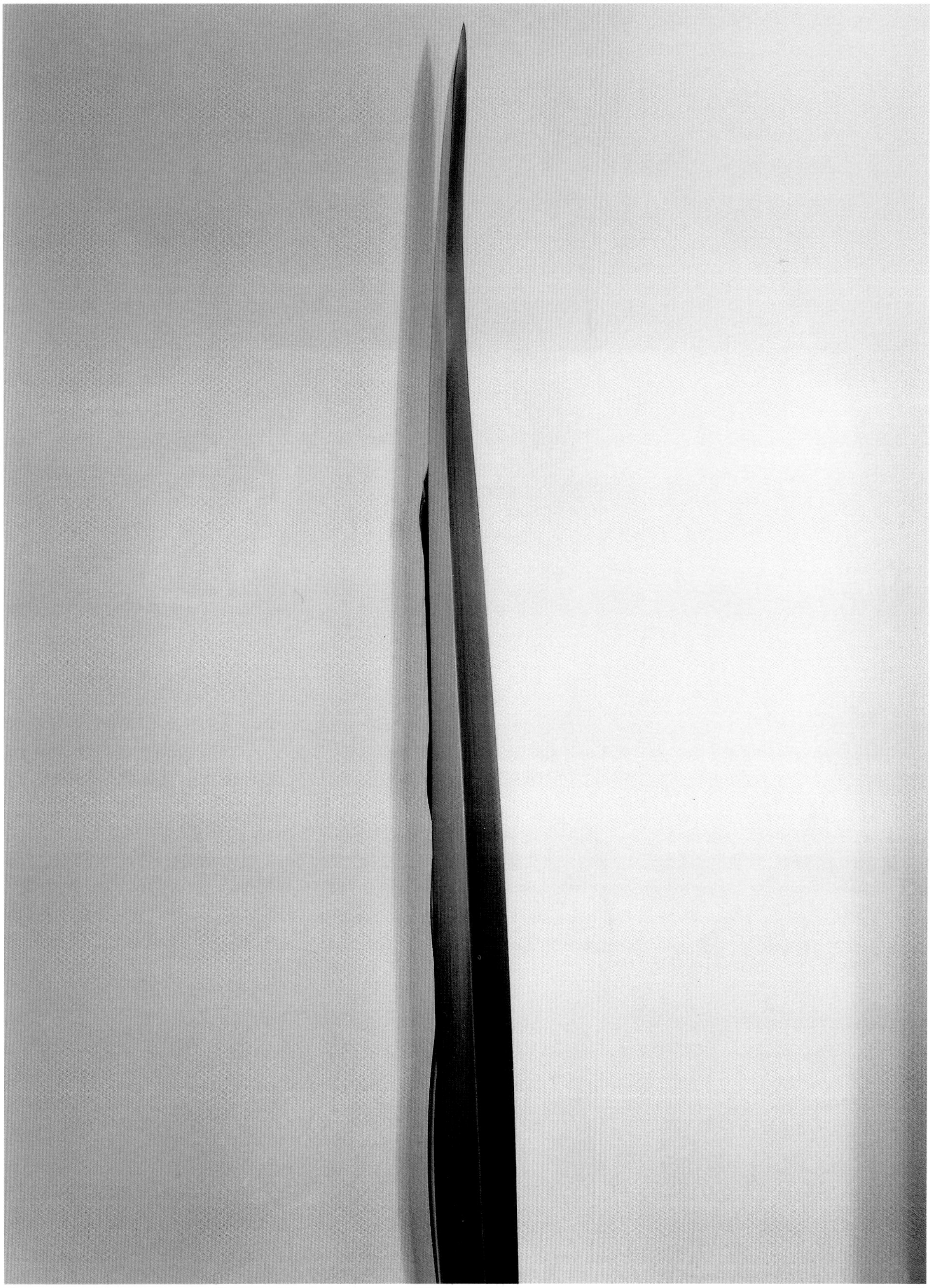

PLATE 7

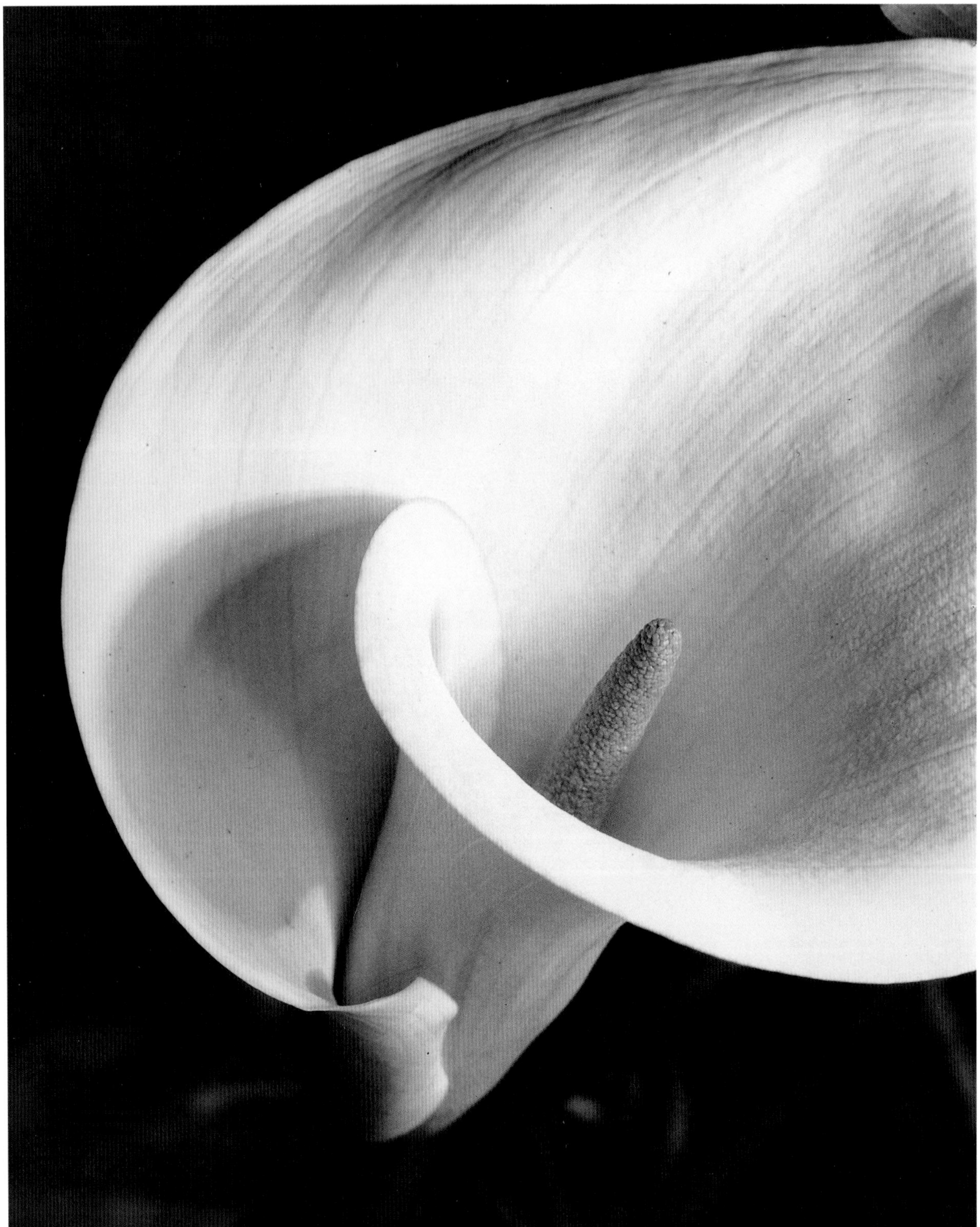

PLATE 8

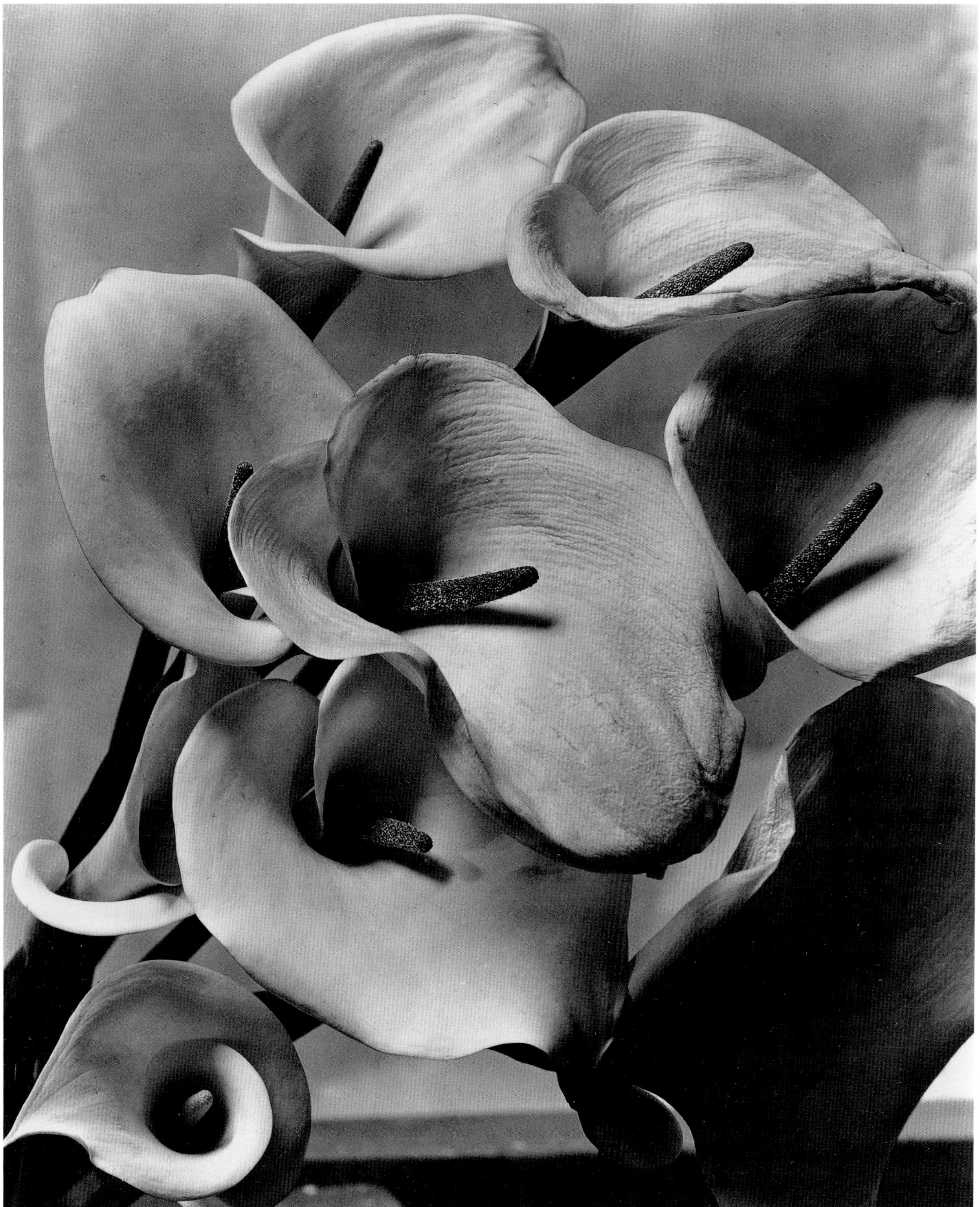

PLATE 9

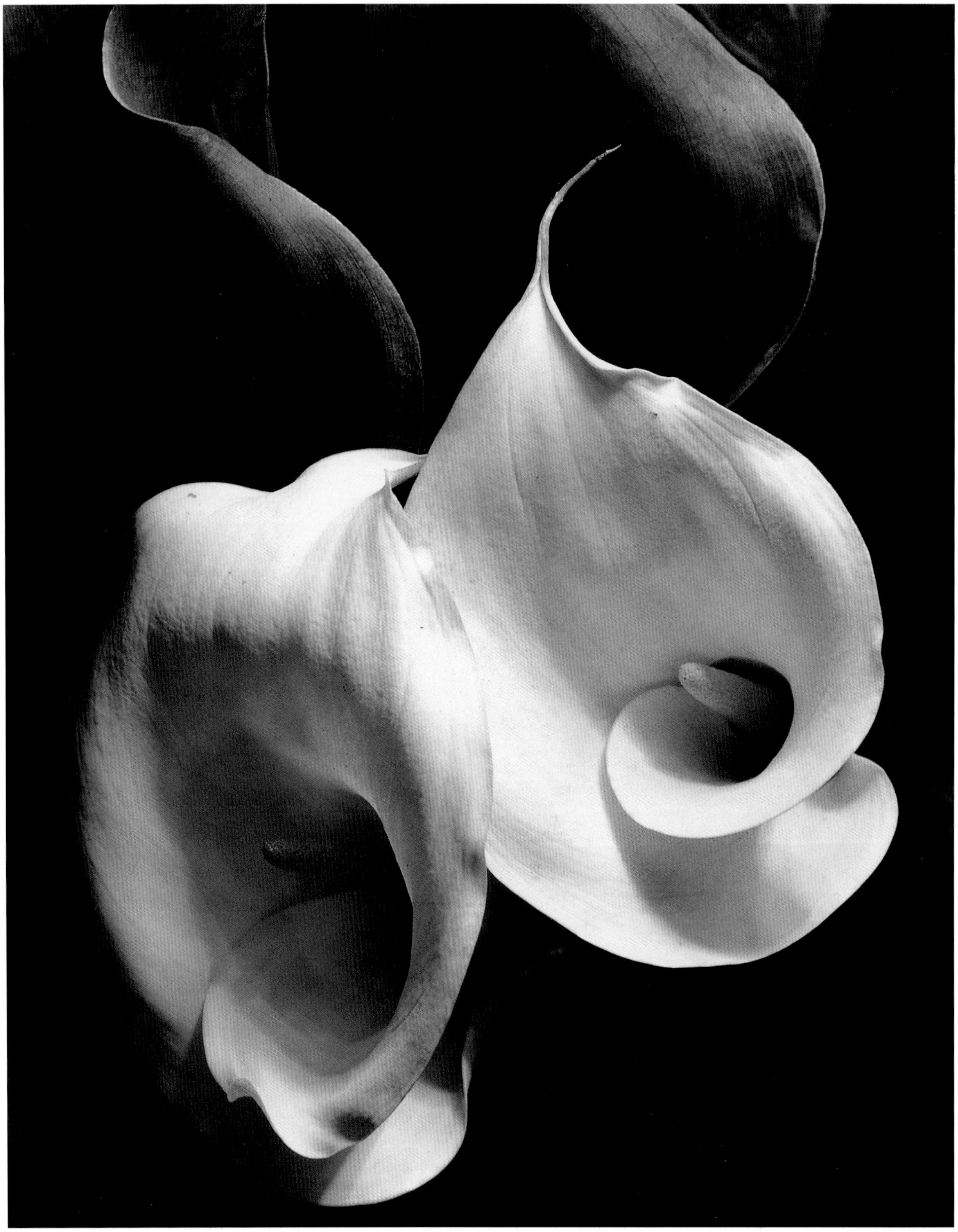

PLATE 10

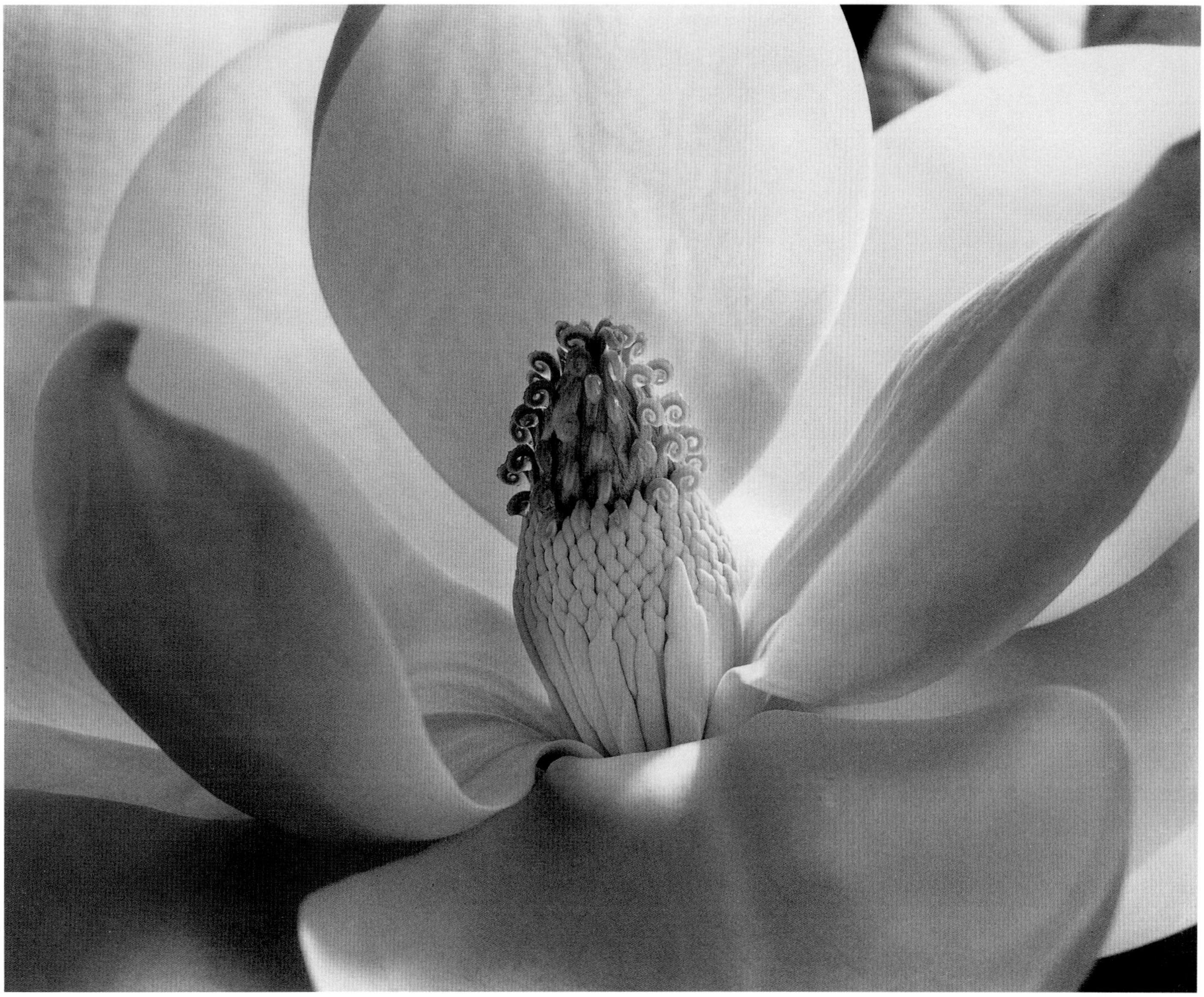

PLATE 11

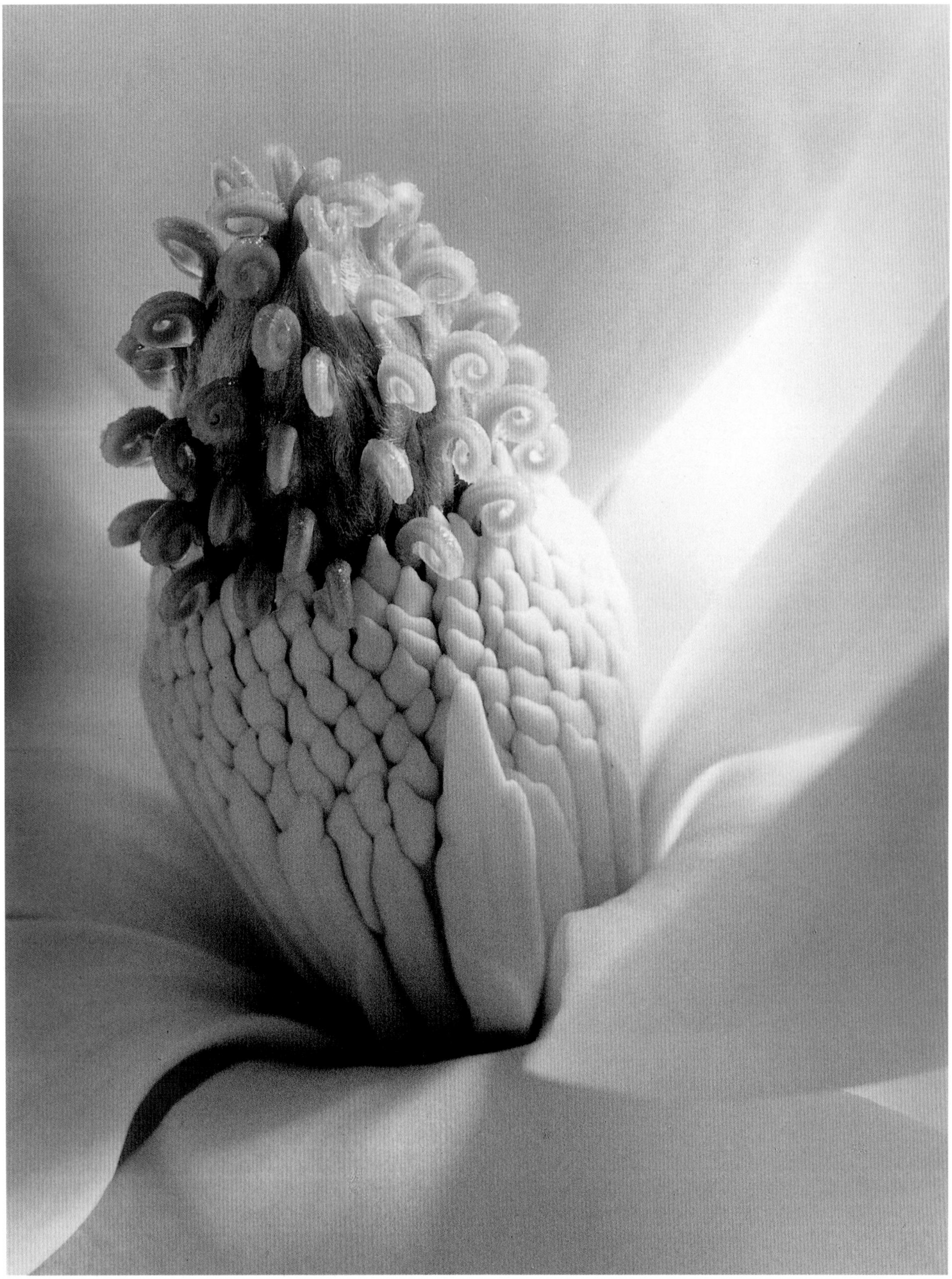

PLATE 12

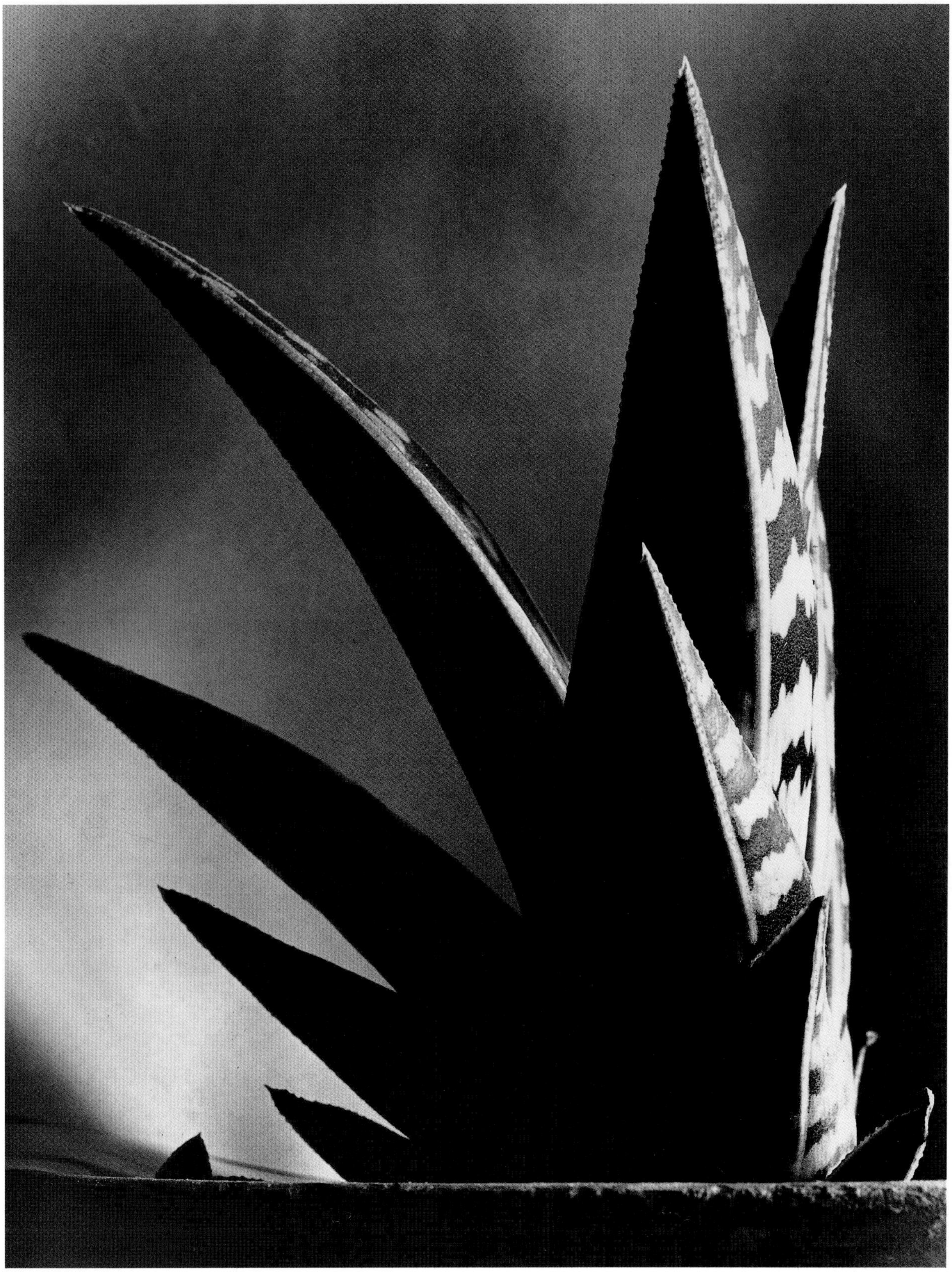

PLATE 13

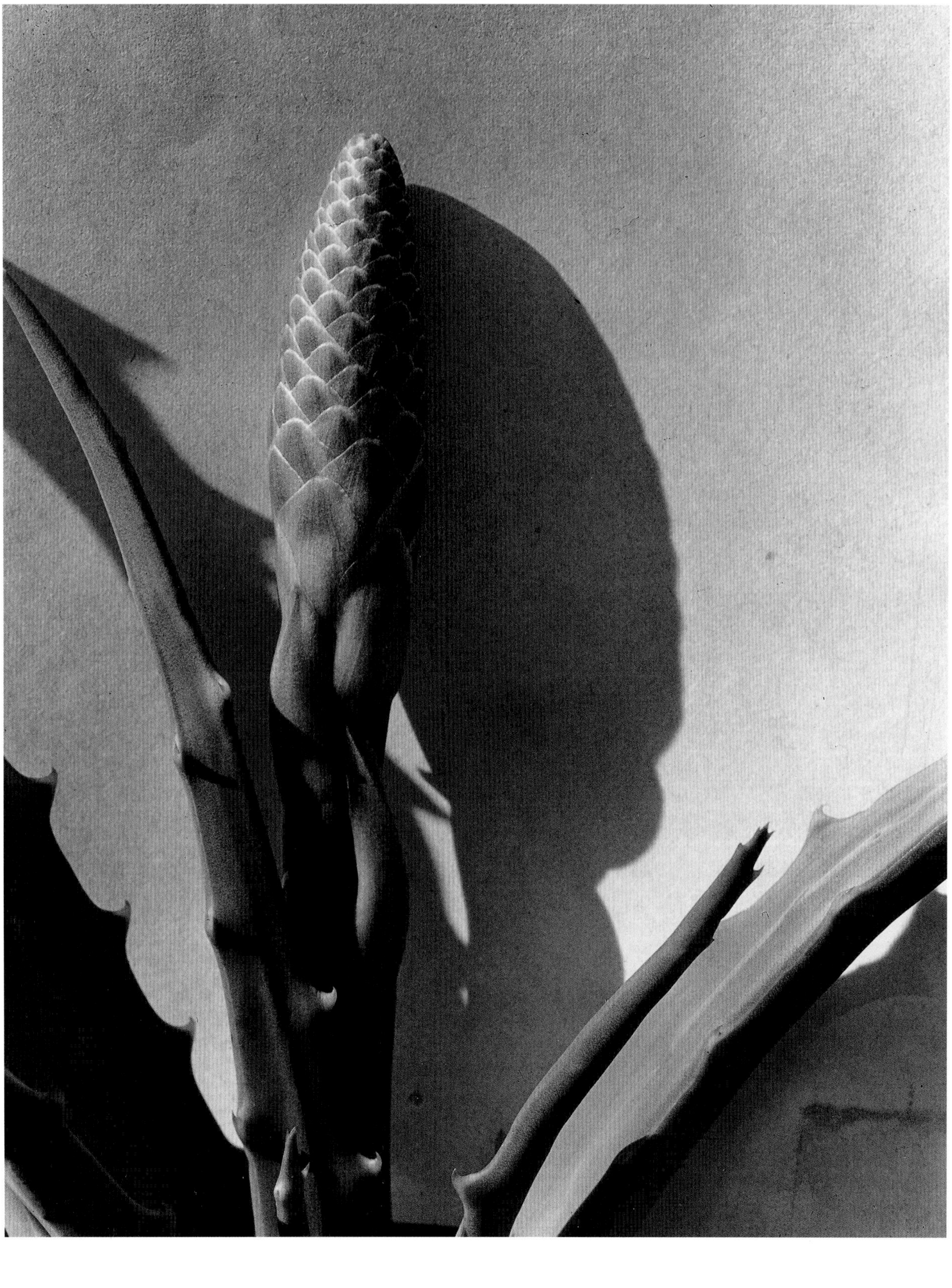

PLATE 14

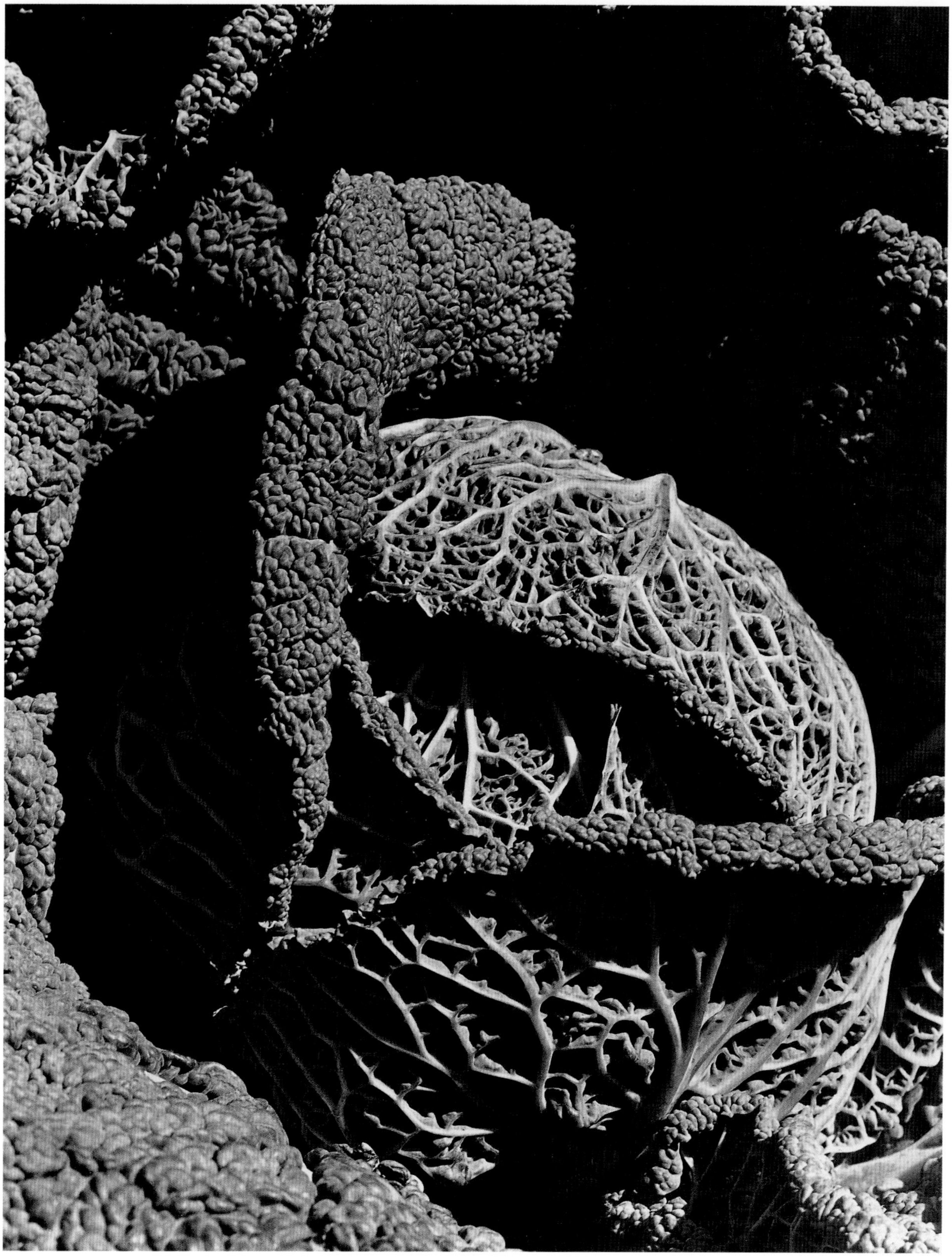

PLATE 15

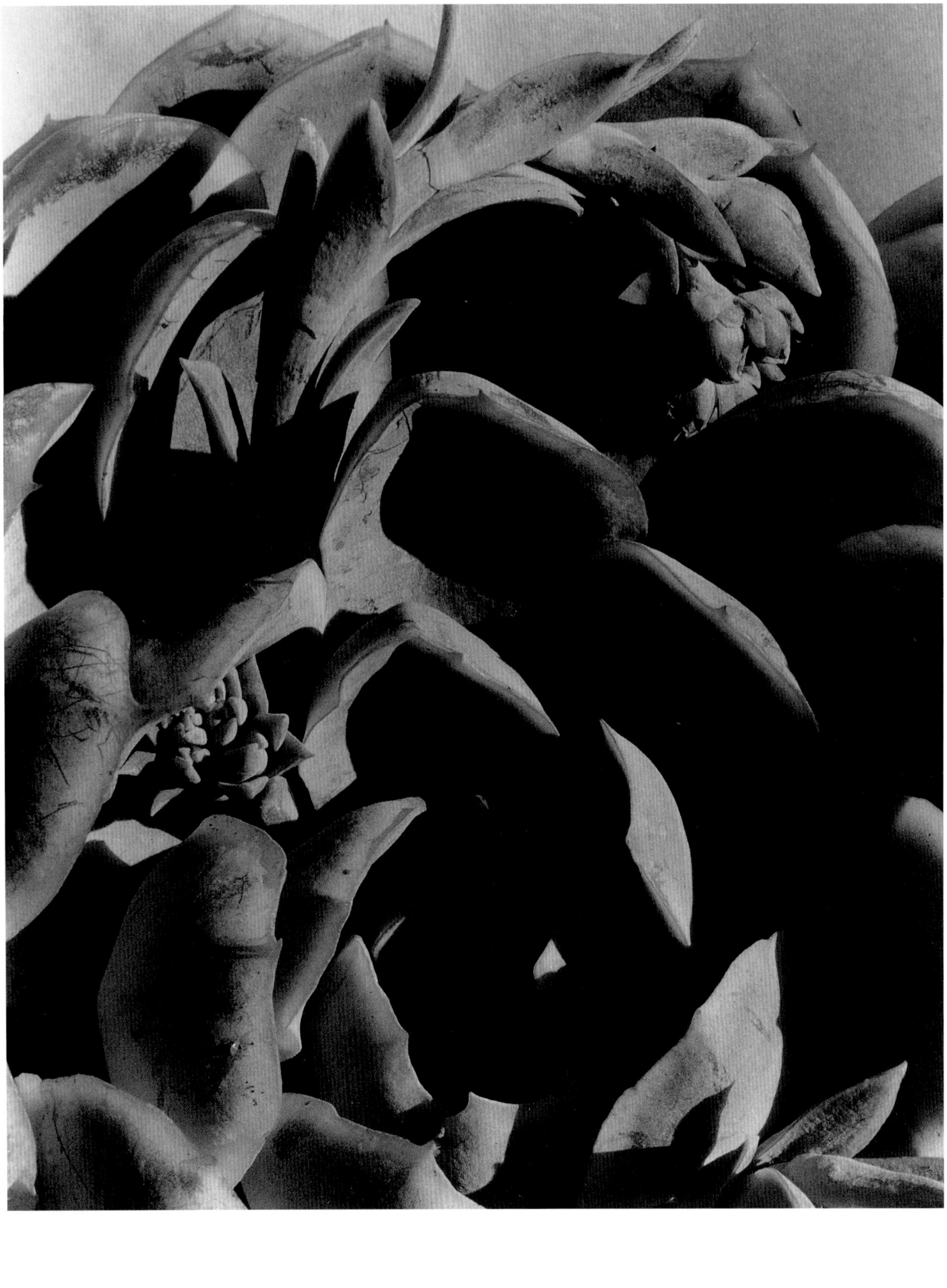

PLATE 16

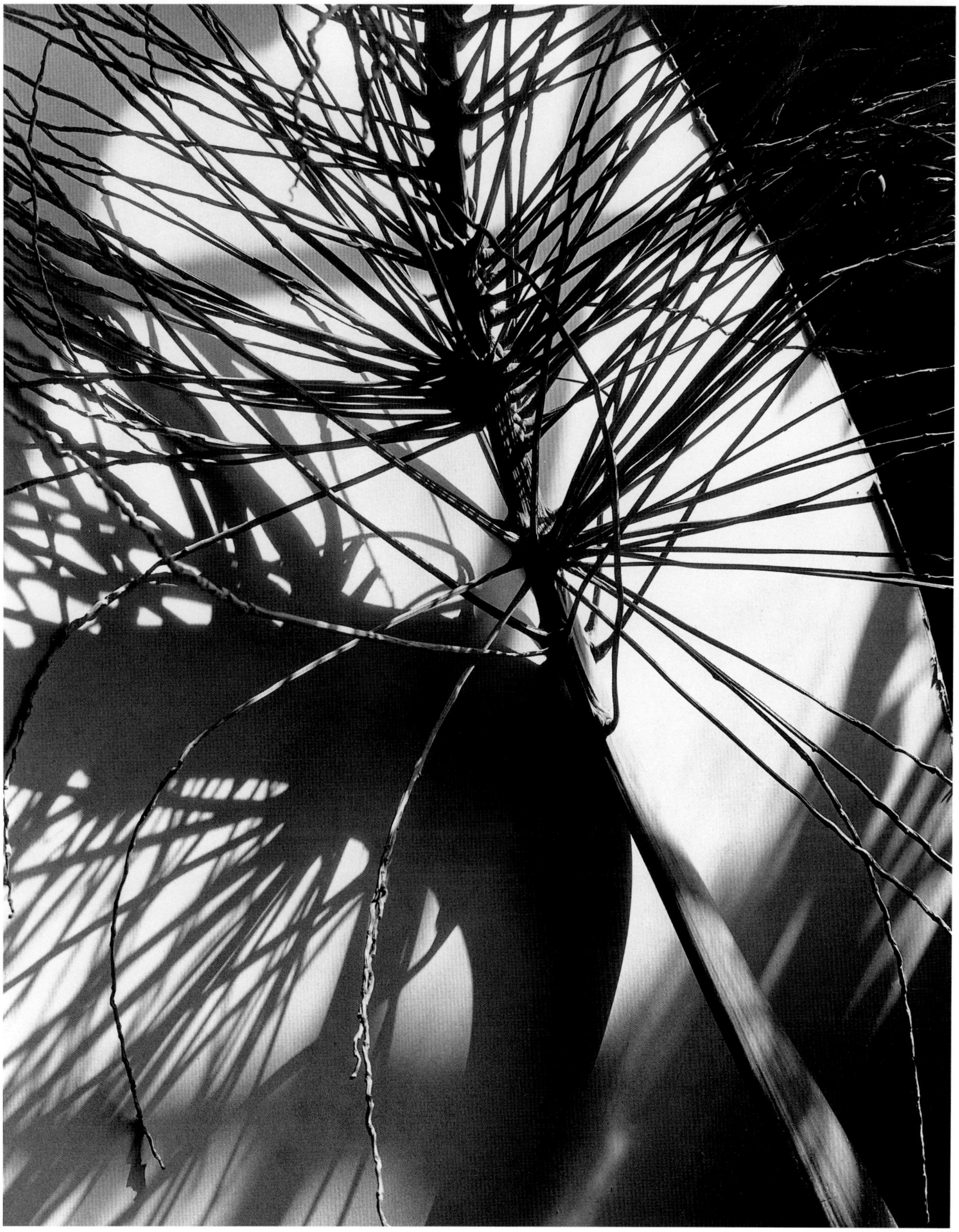

PLATE 17

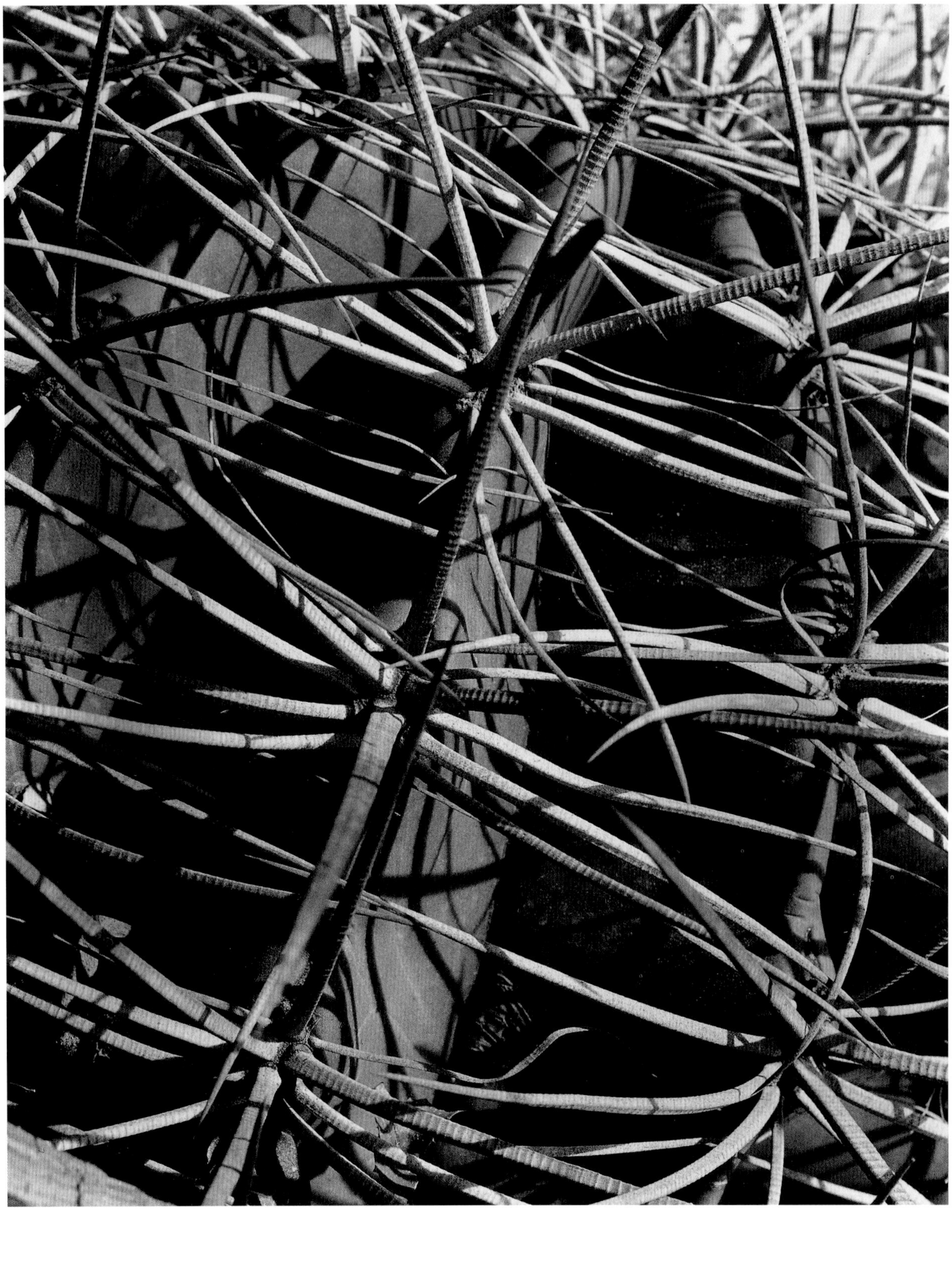

PLATE 18

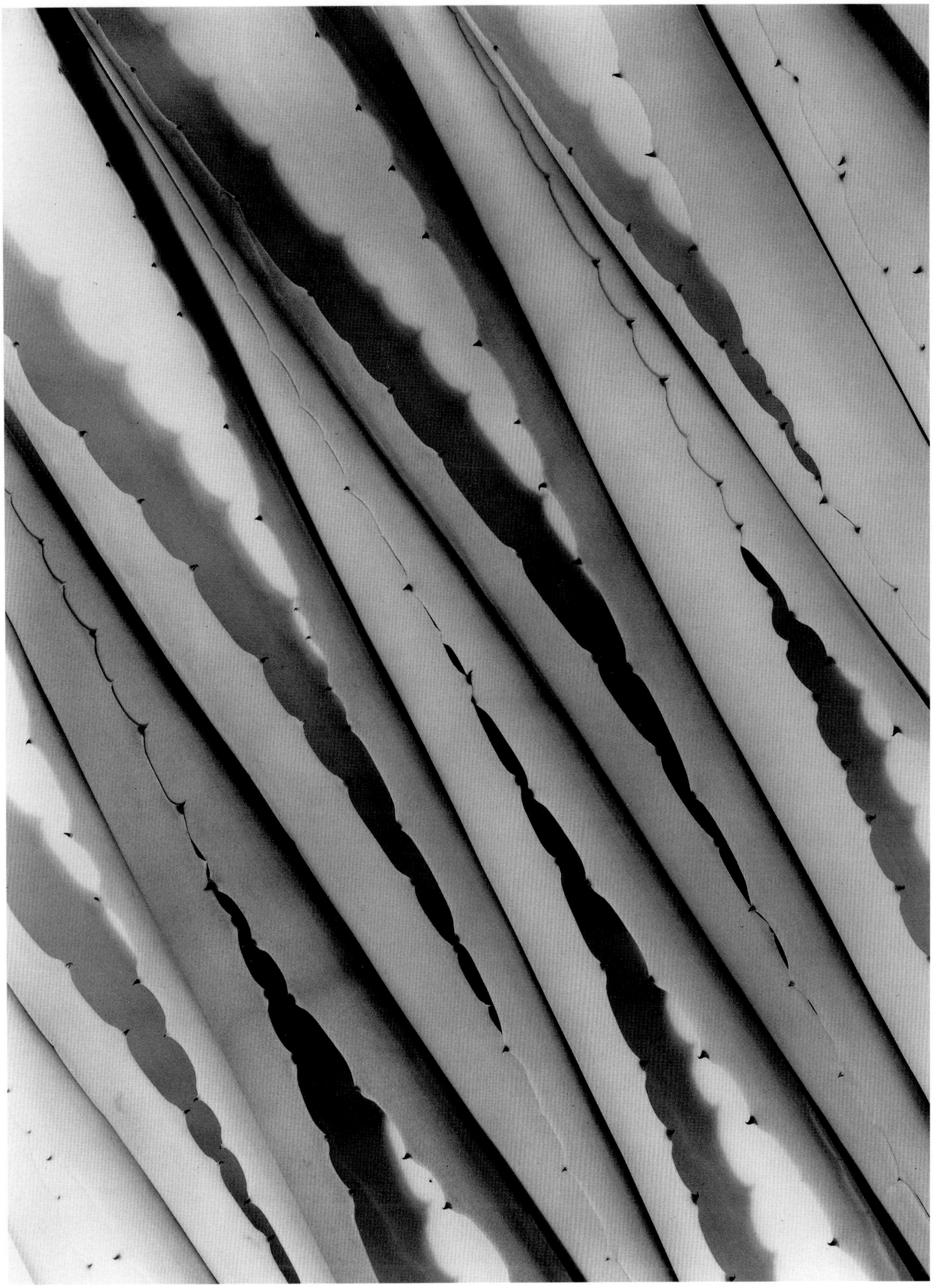

PLATE 19

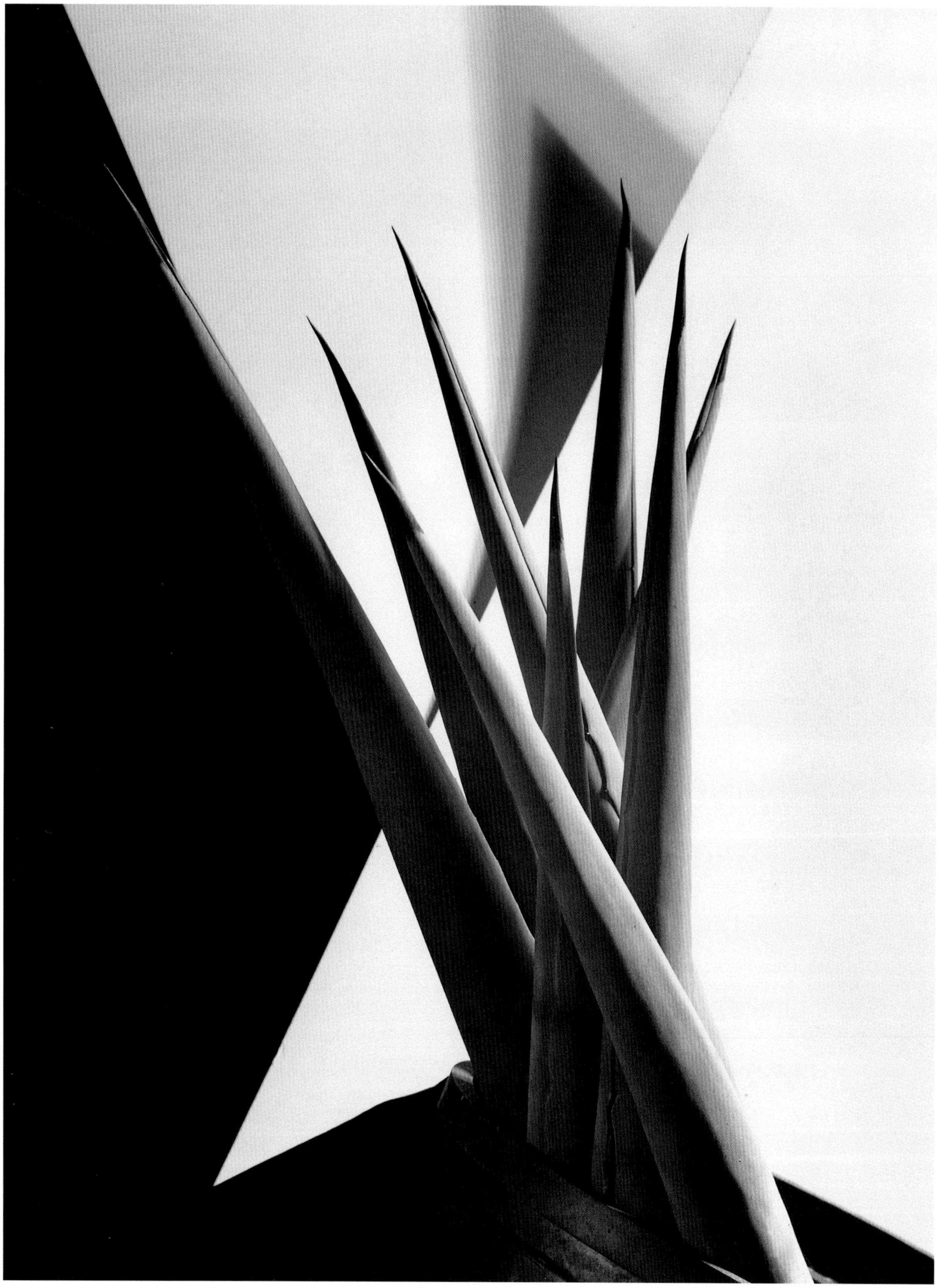

PLATE 20

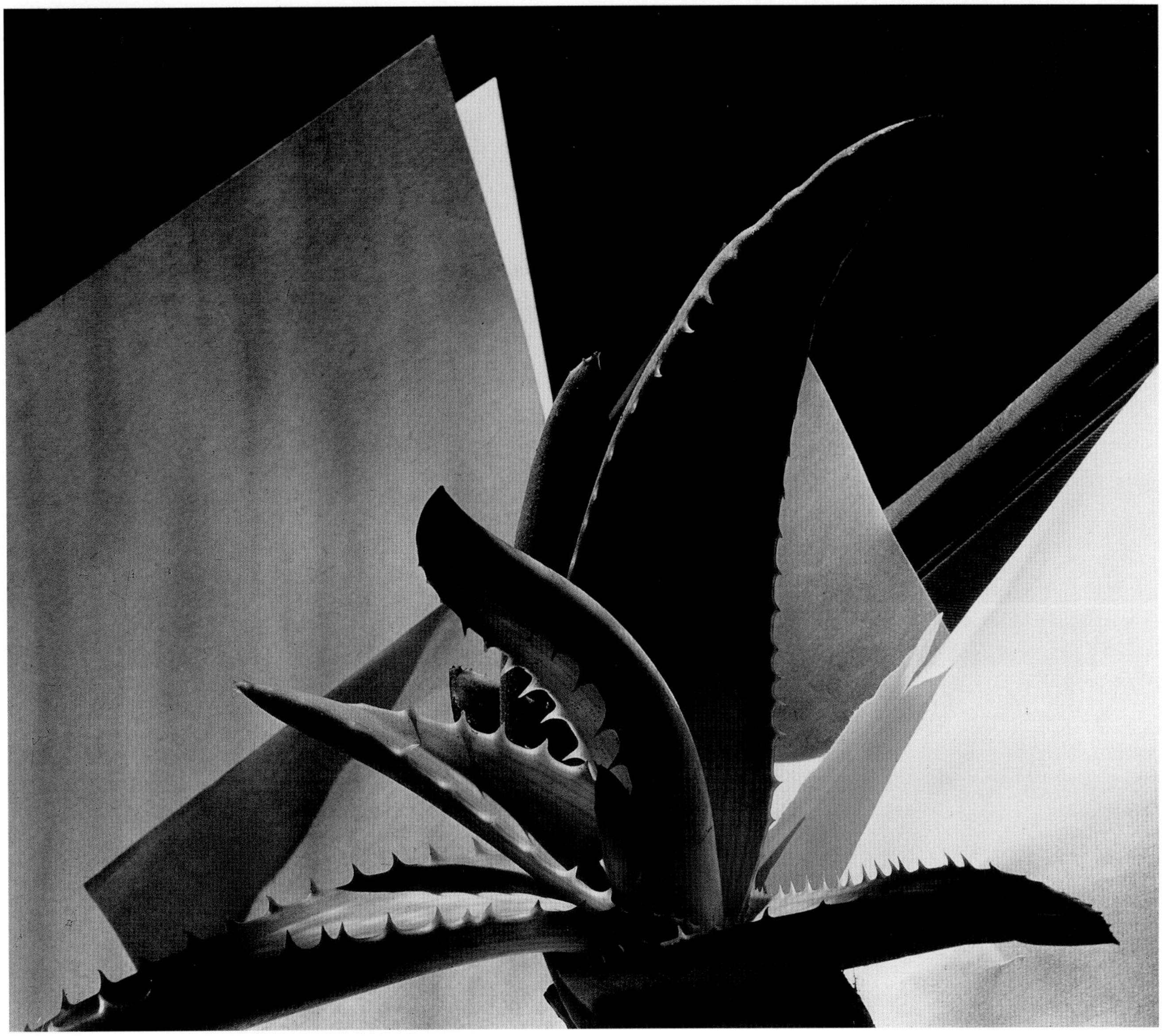

PLATE 21

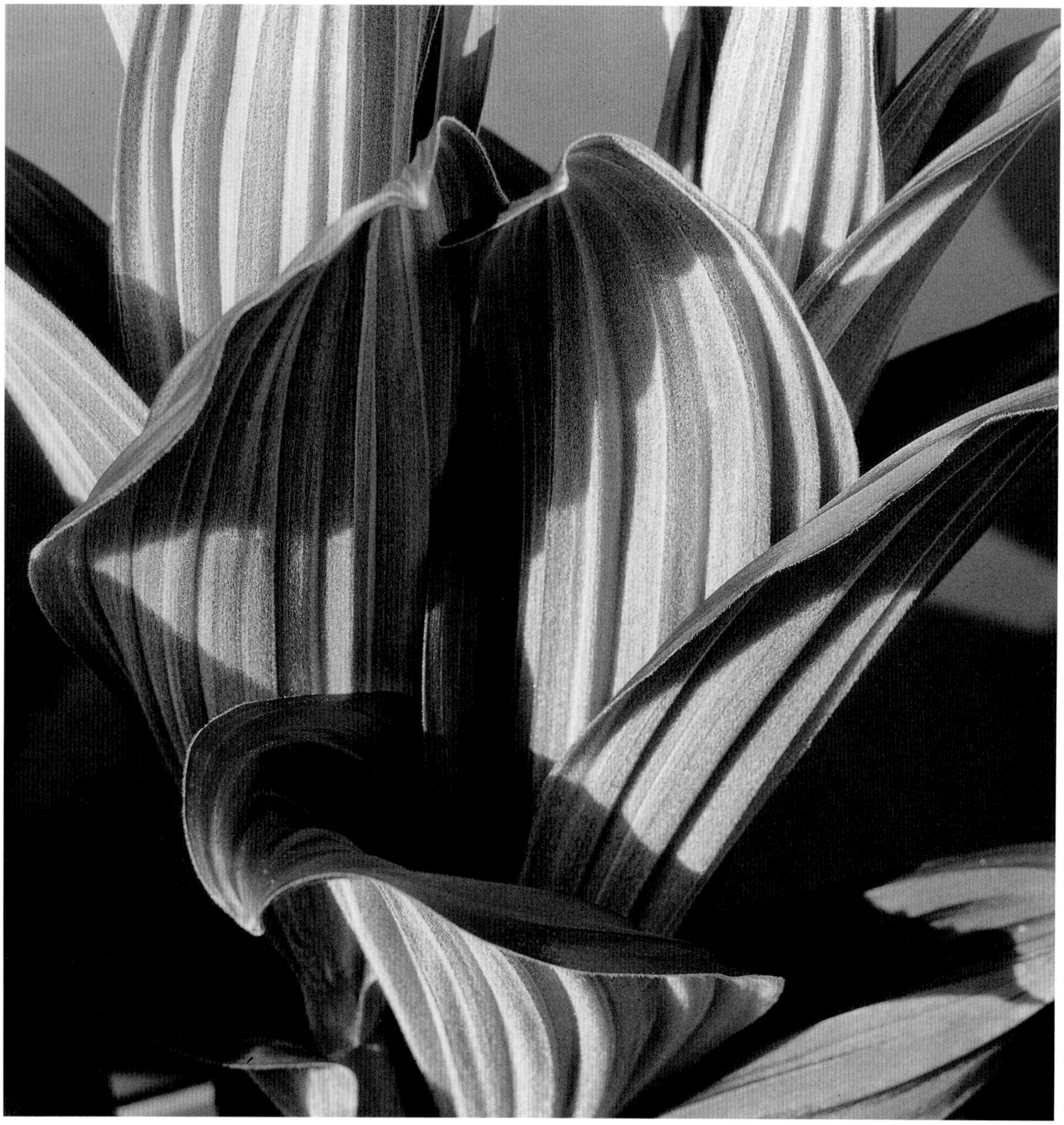

PLATE 22

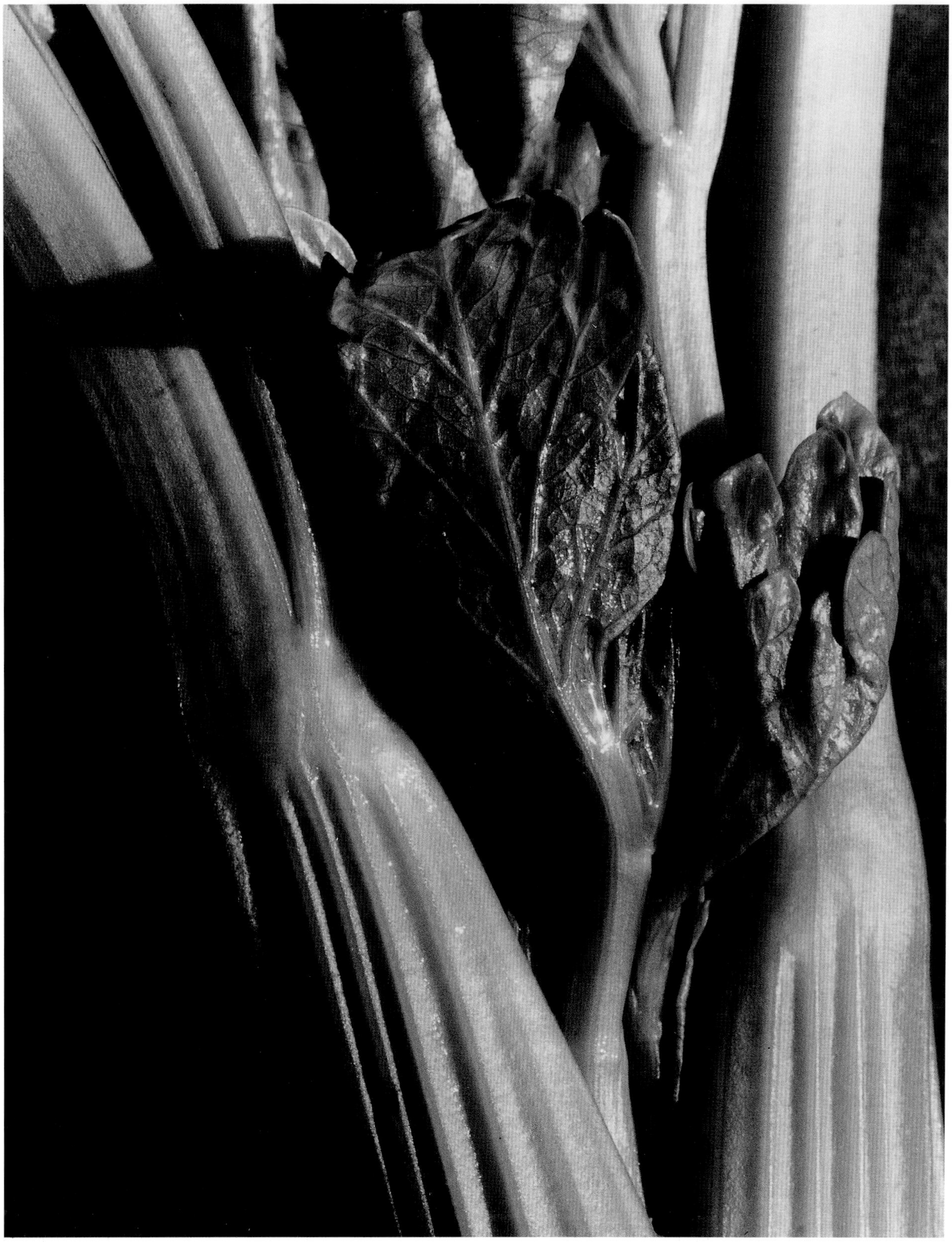

PLATE 23

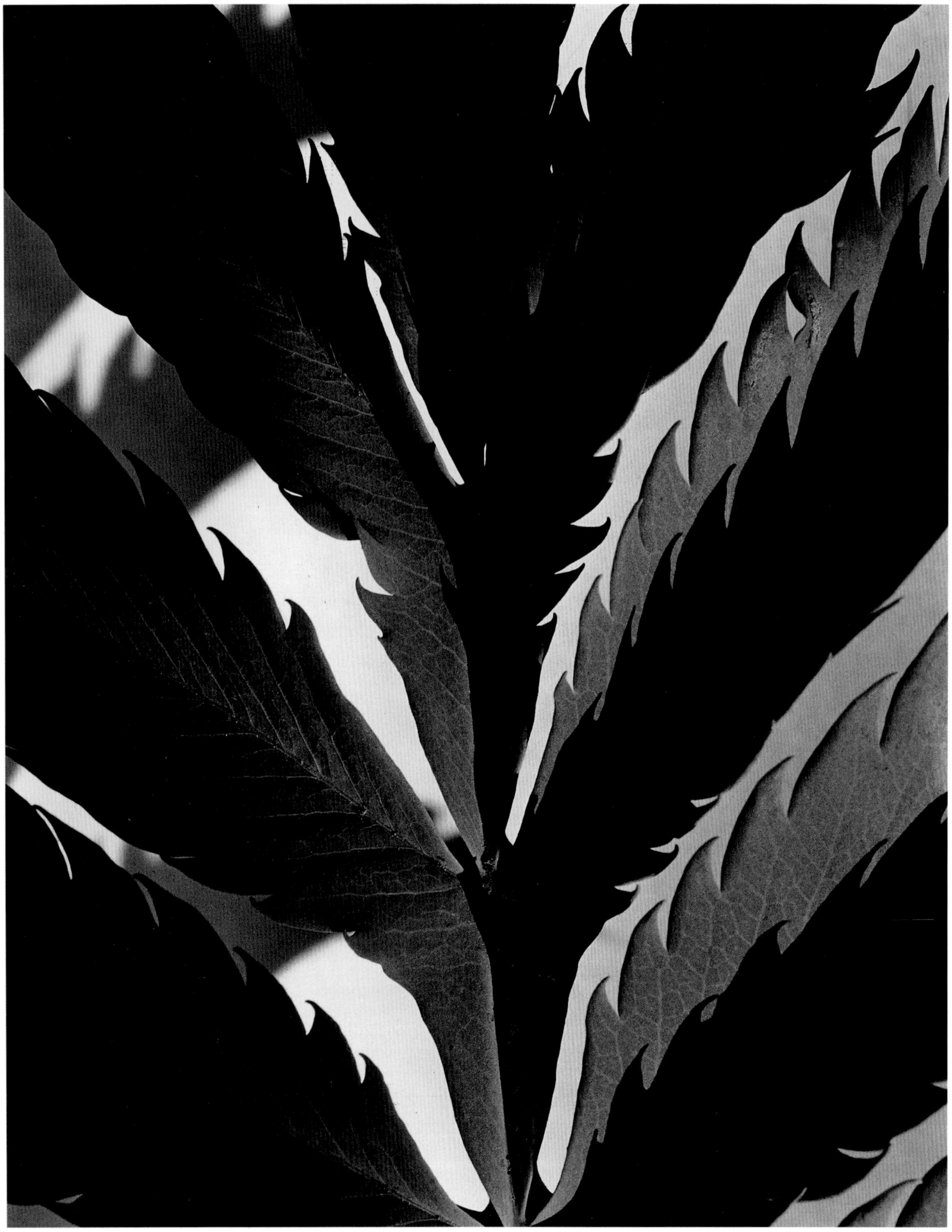

PLATE 24

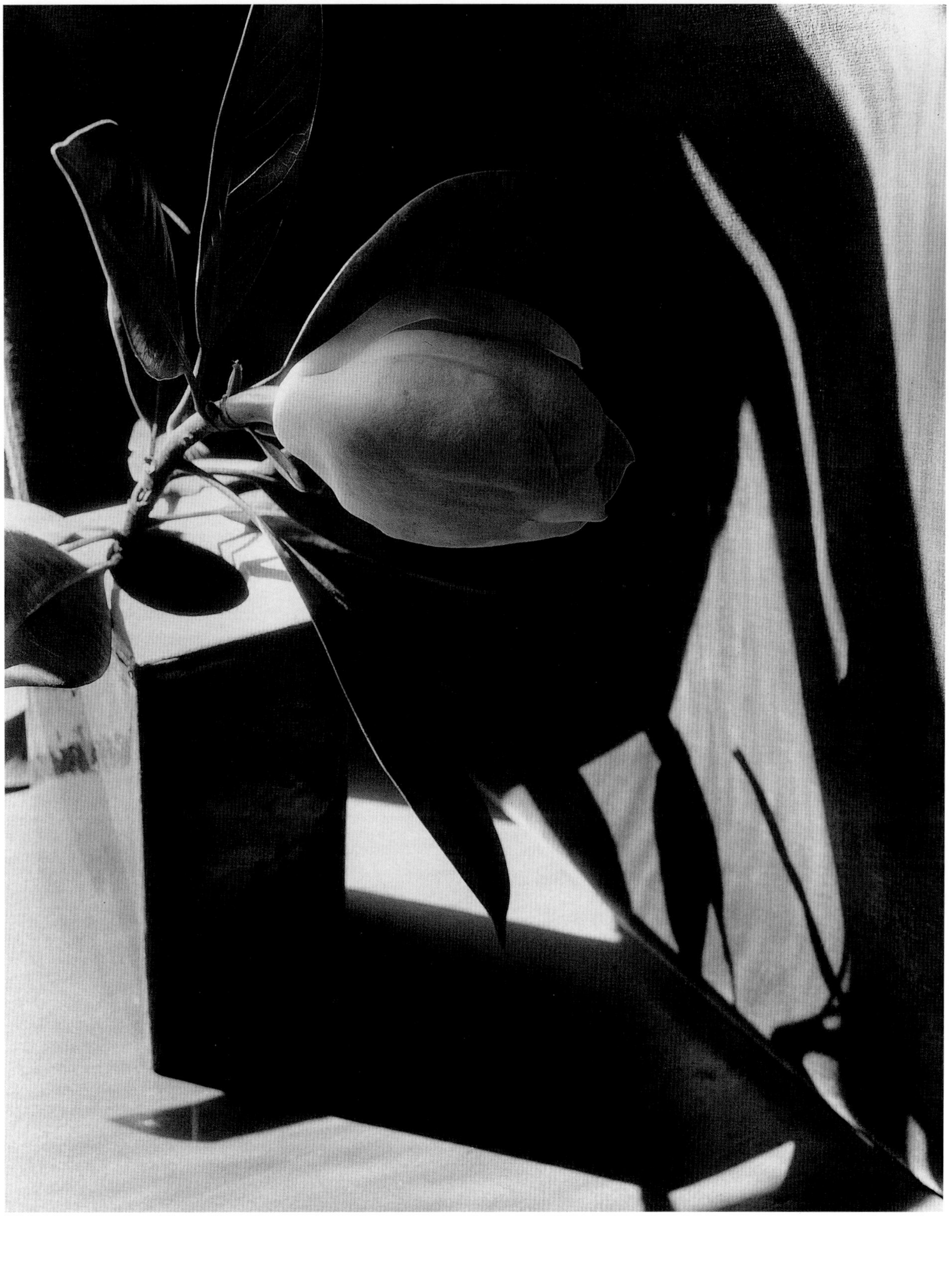

PLATE 25

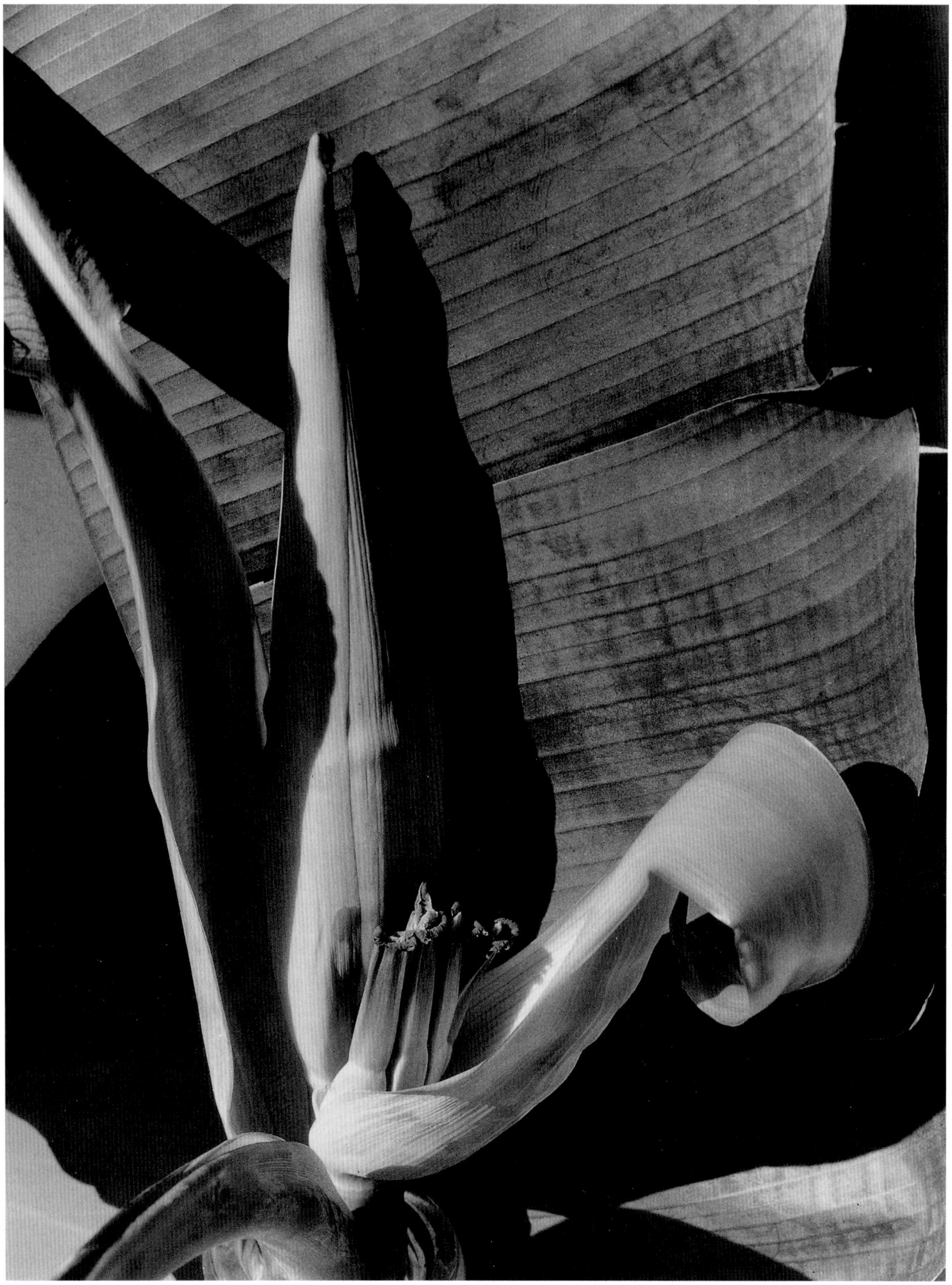

PLATE 26

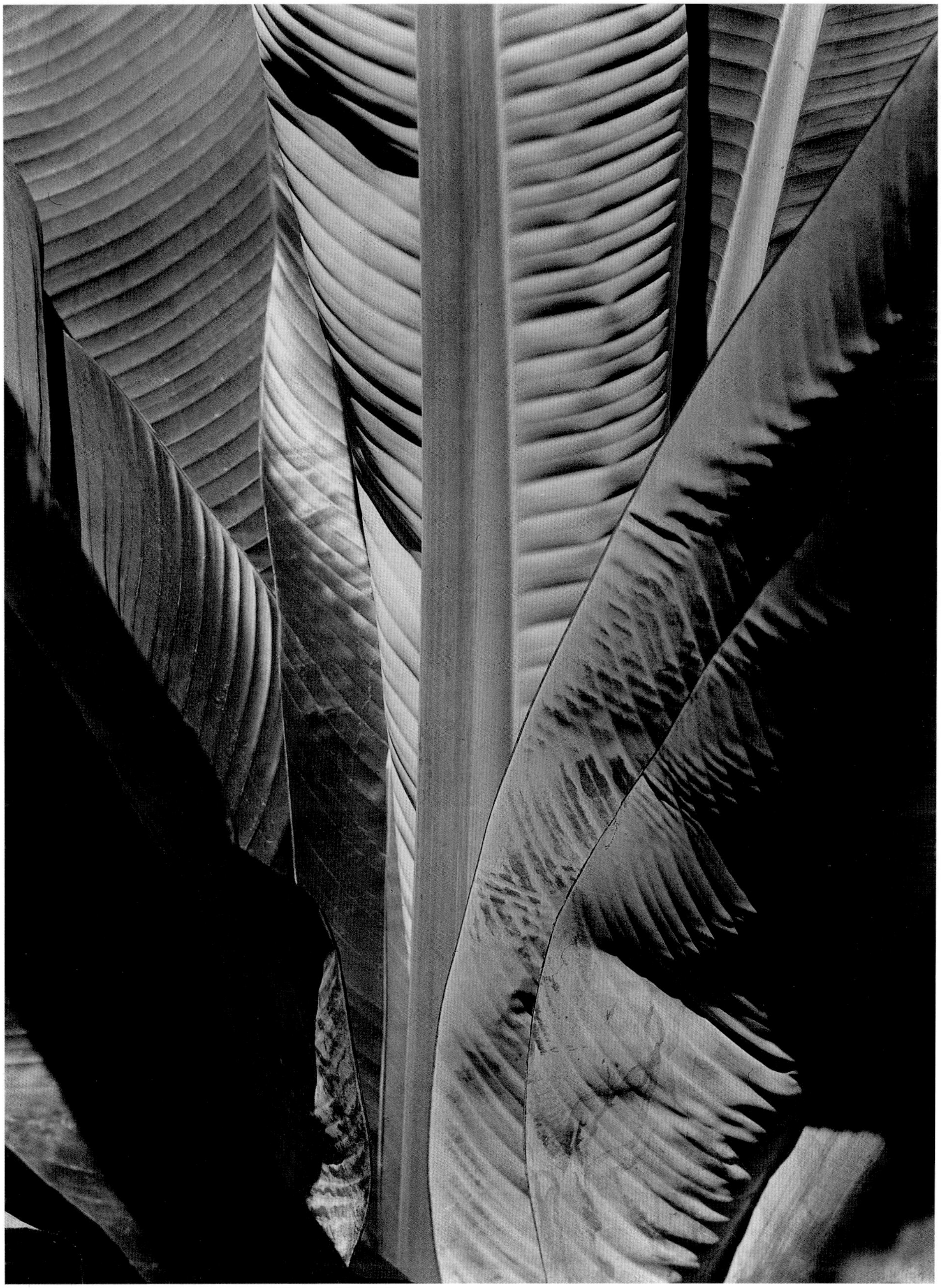

PLATE 27

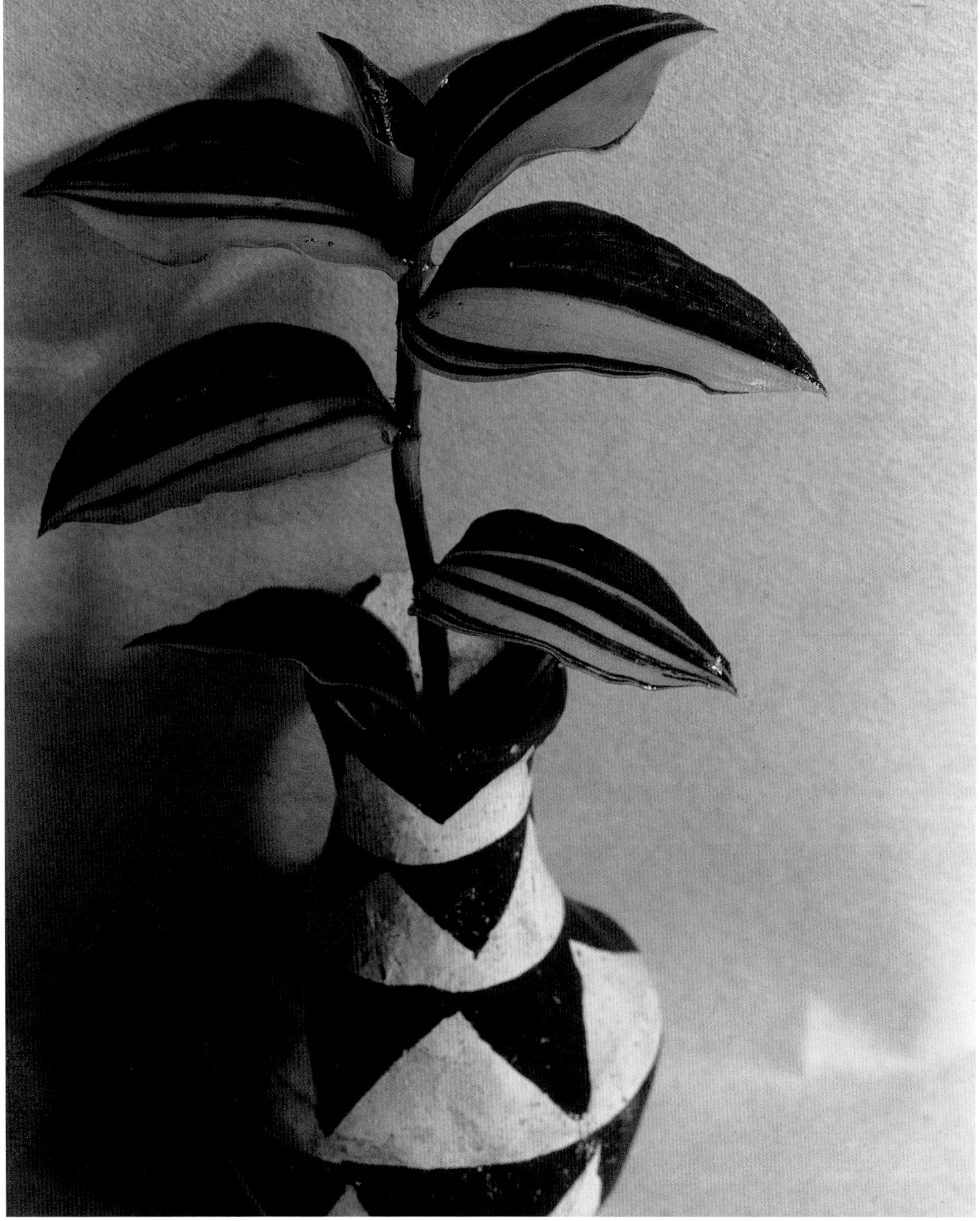

PLATE 28

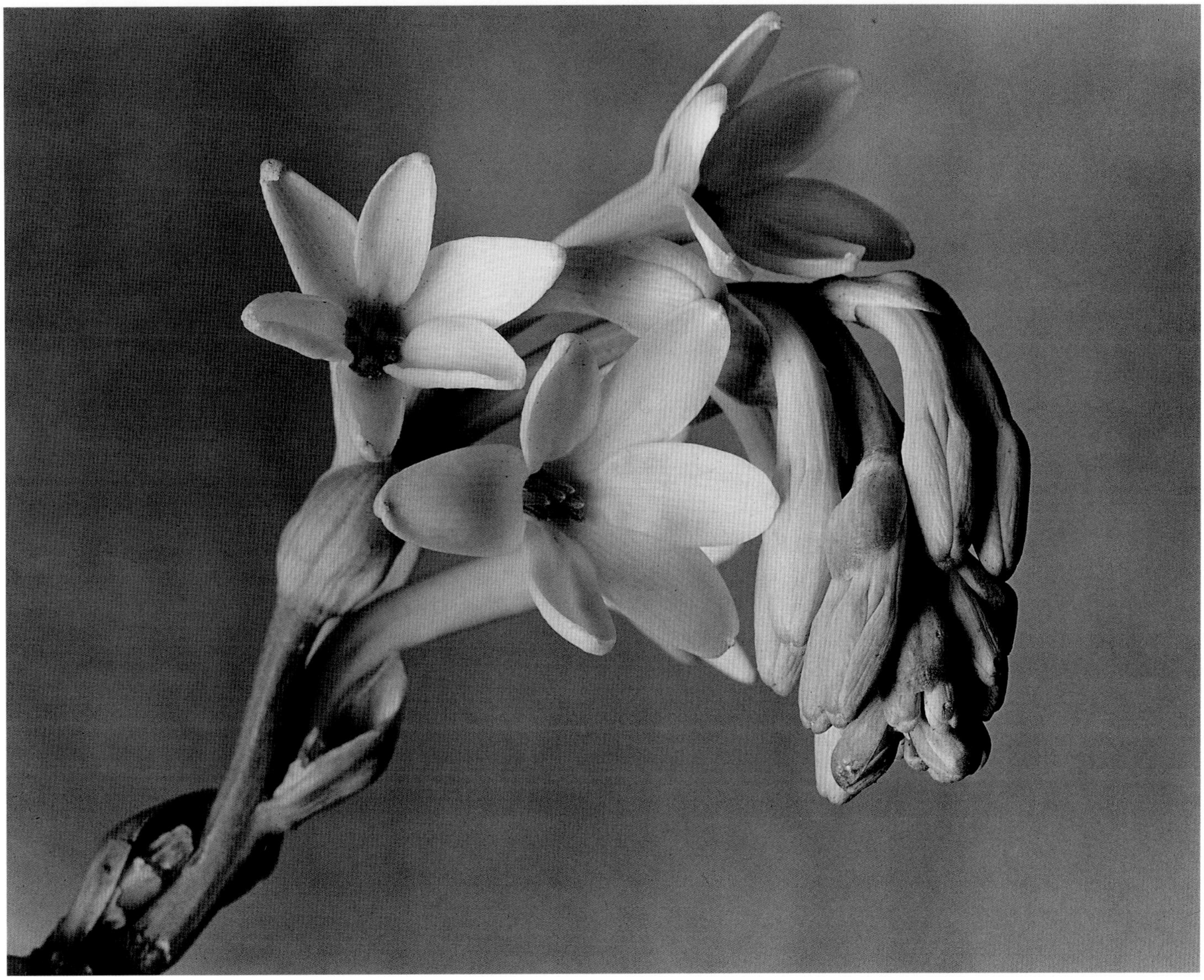

PLATE 29

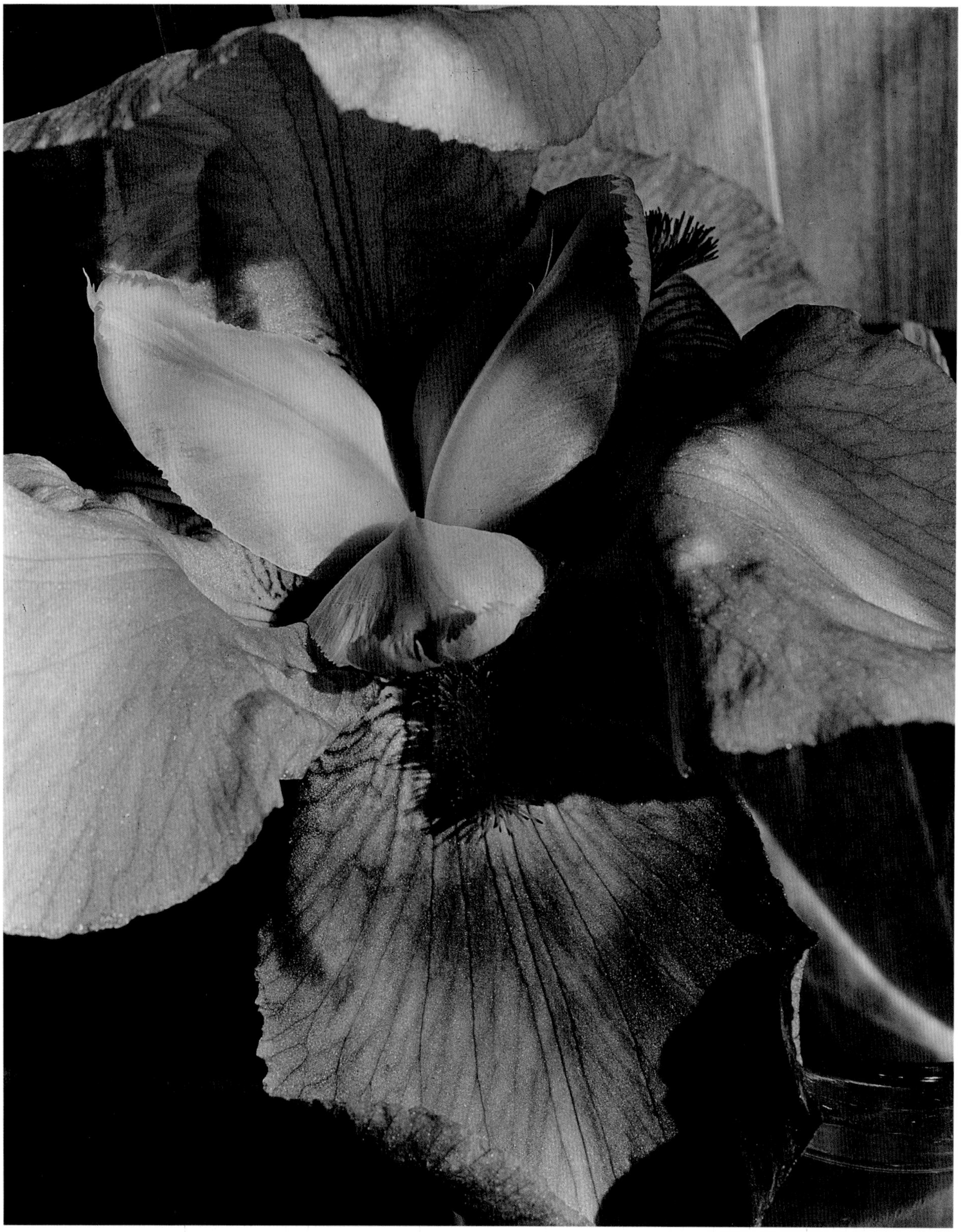

PLATE 30

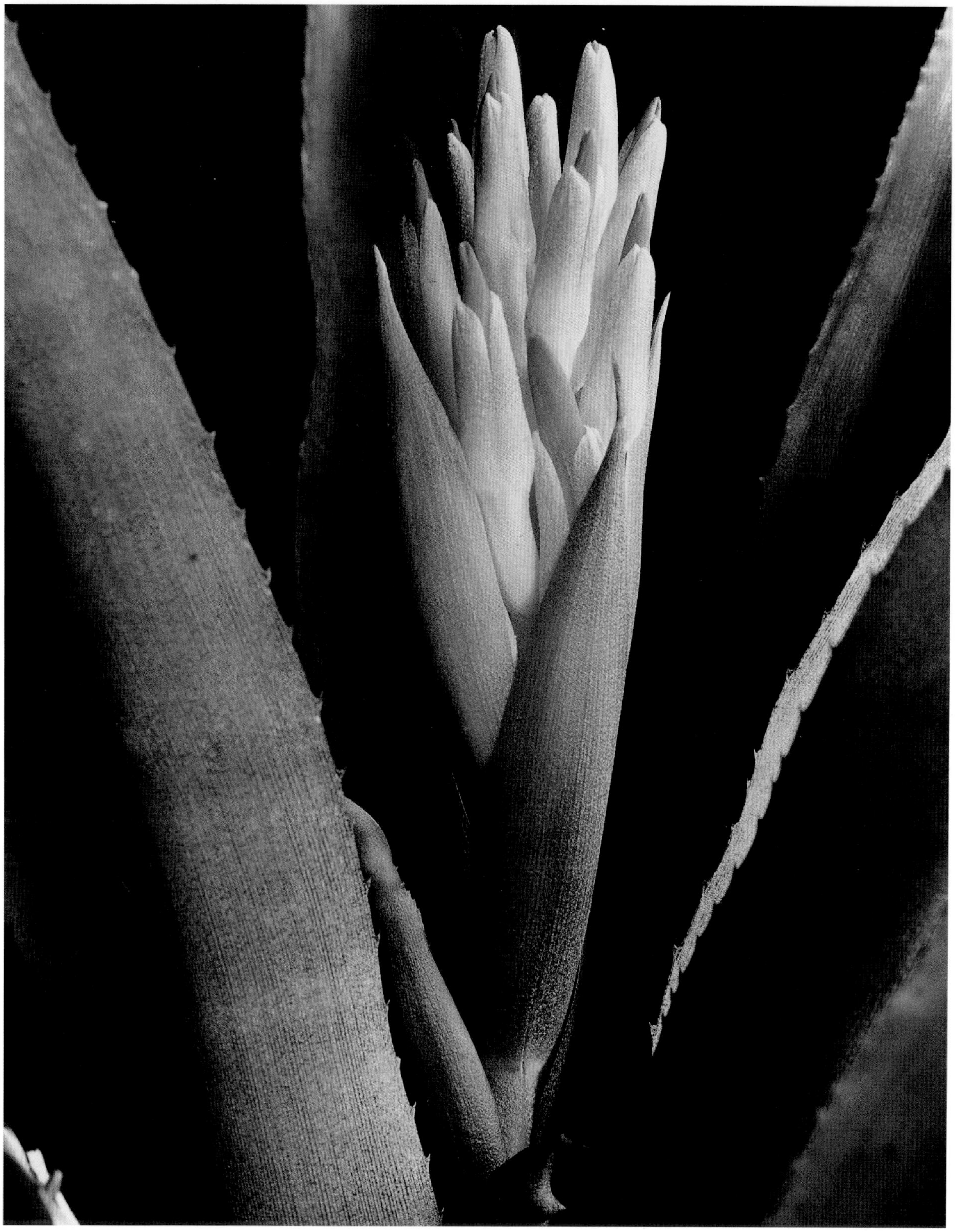

PLATE 31

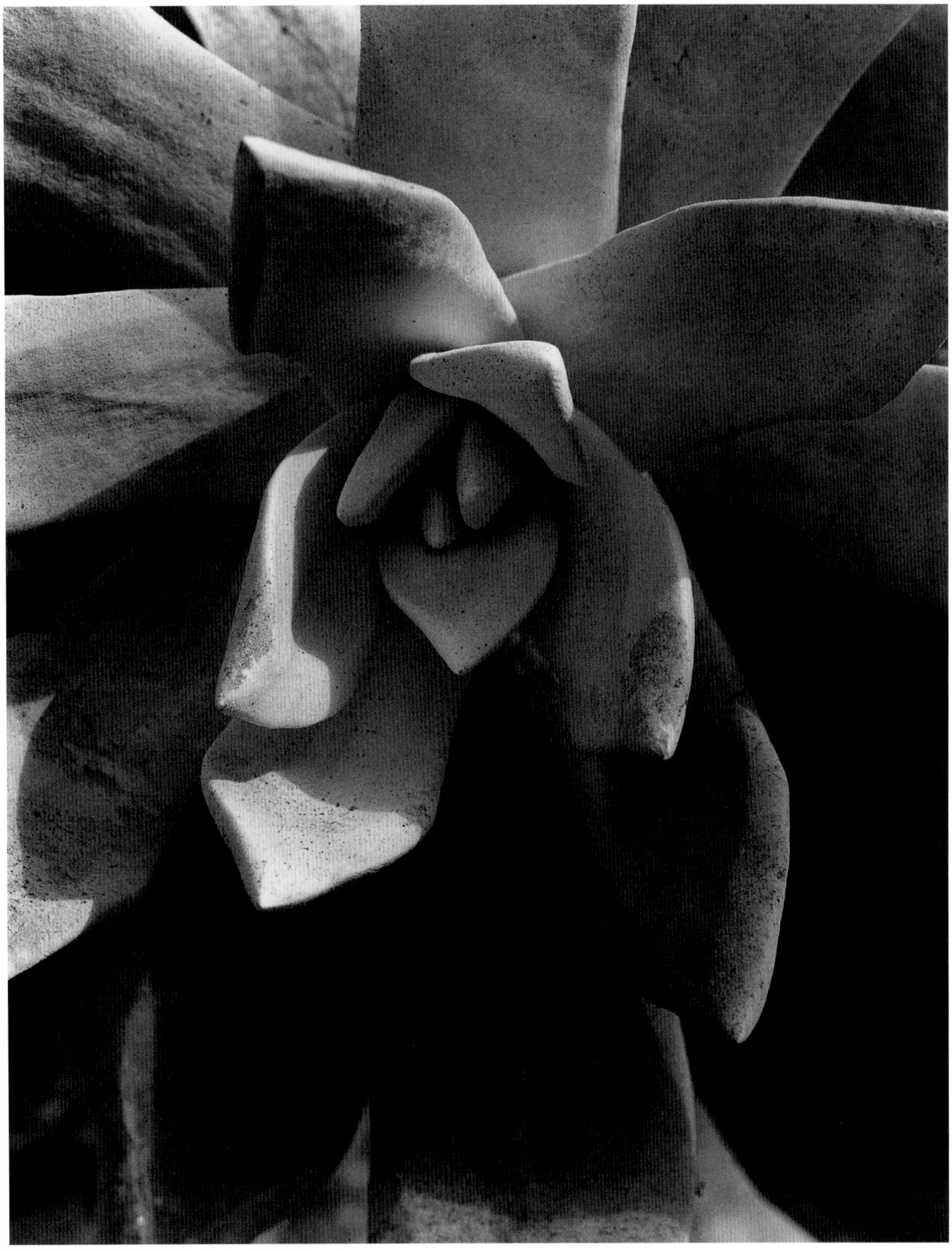

PLATE 32

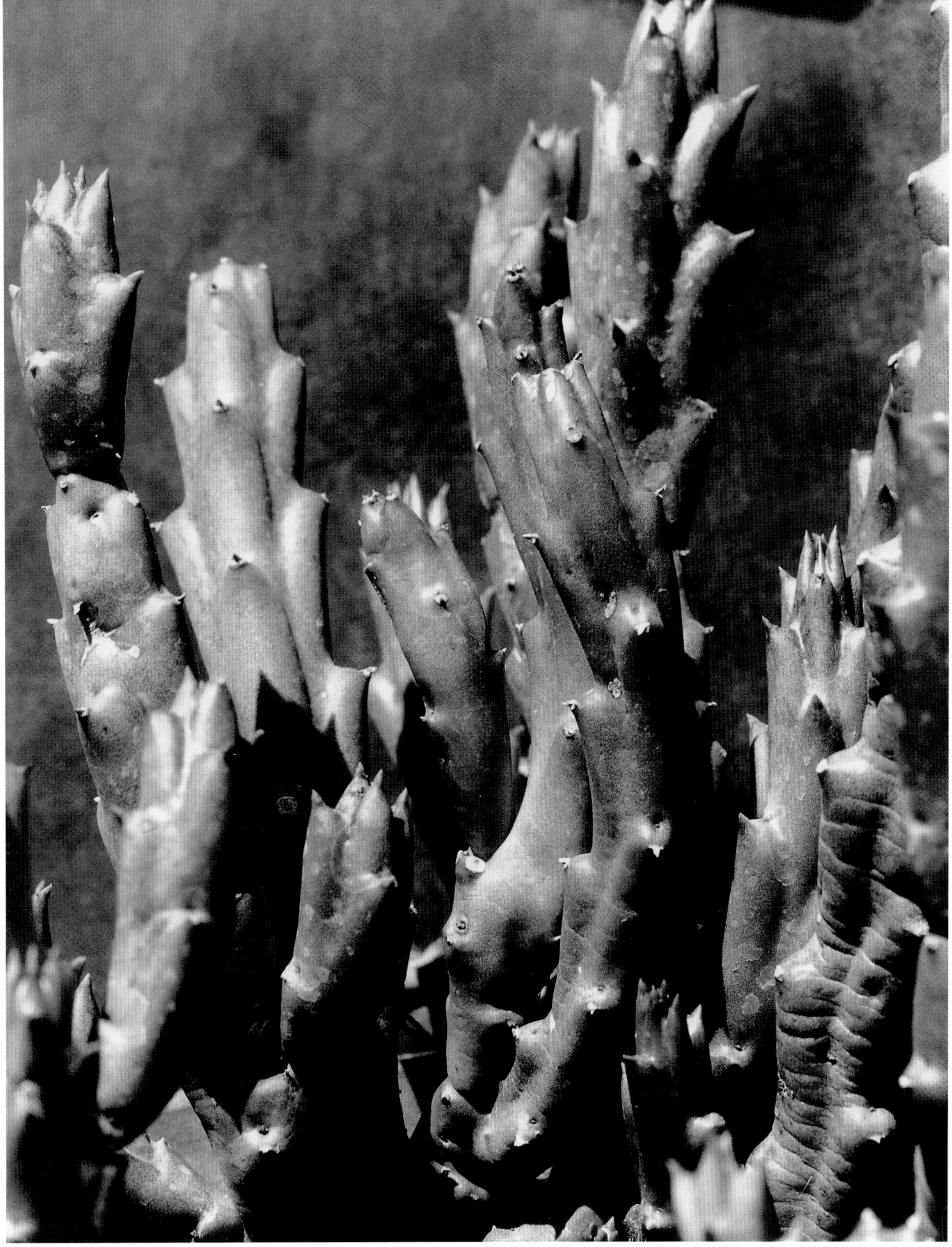

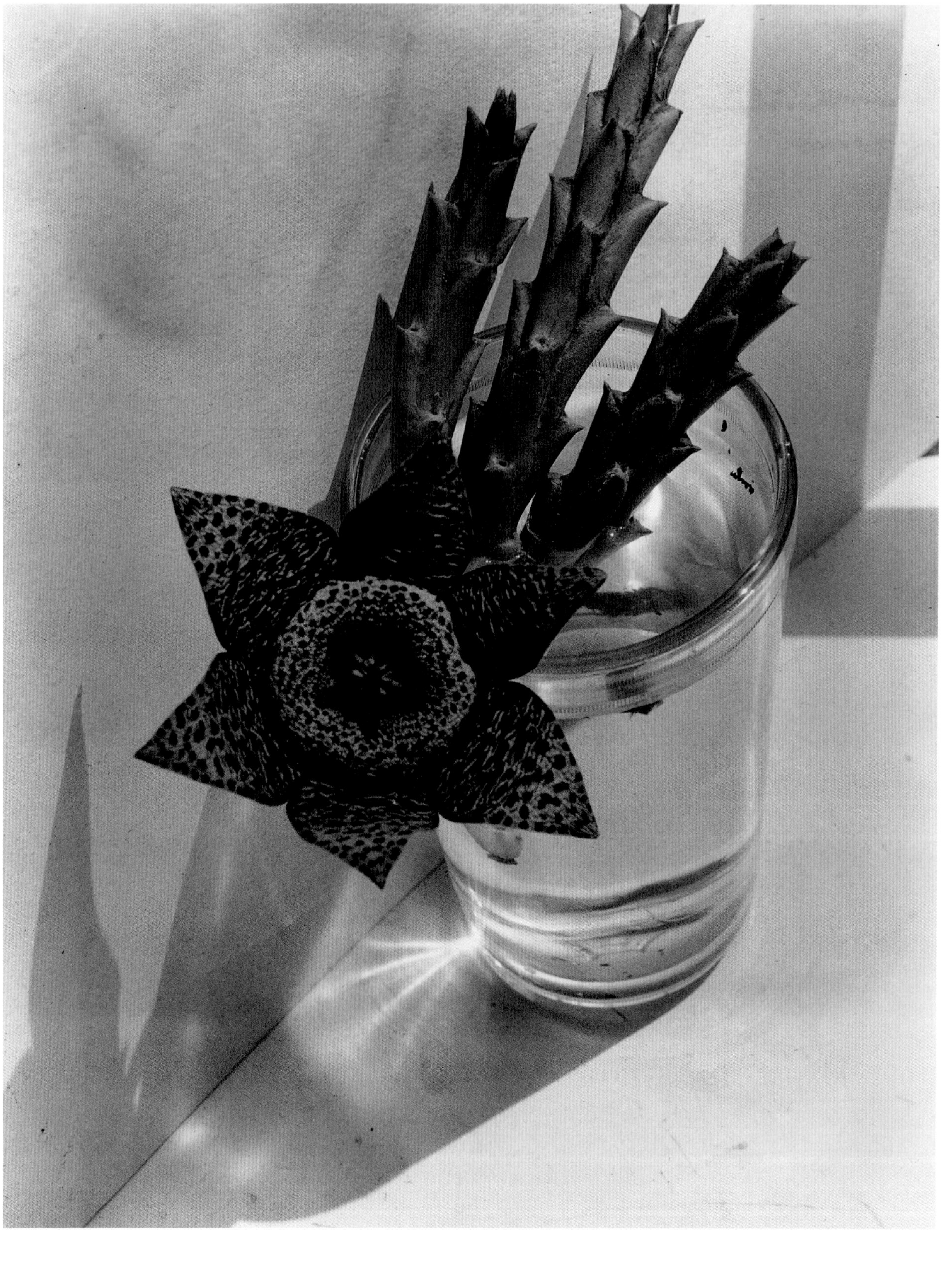

PLATE 34

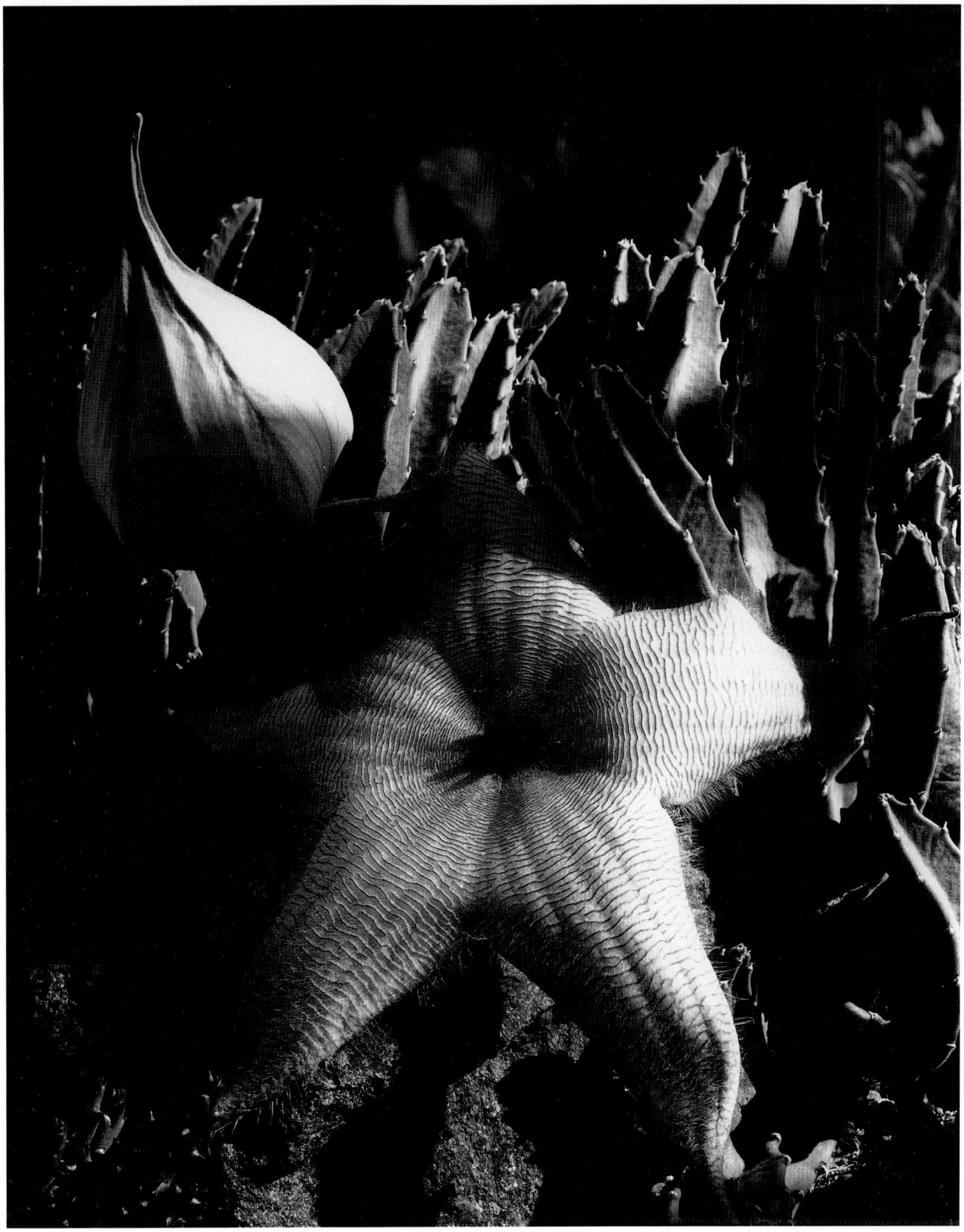

PLATE 35

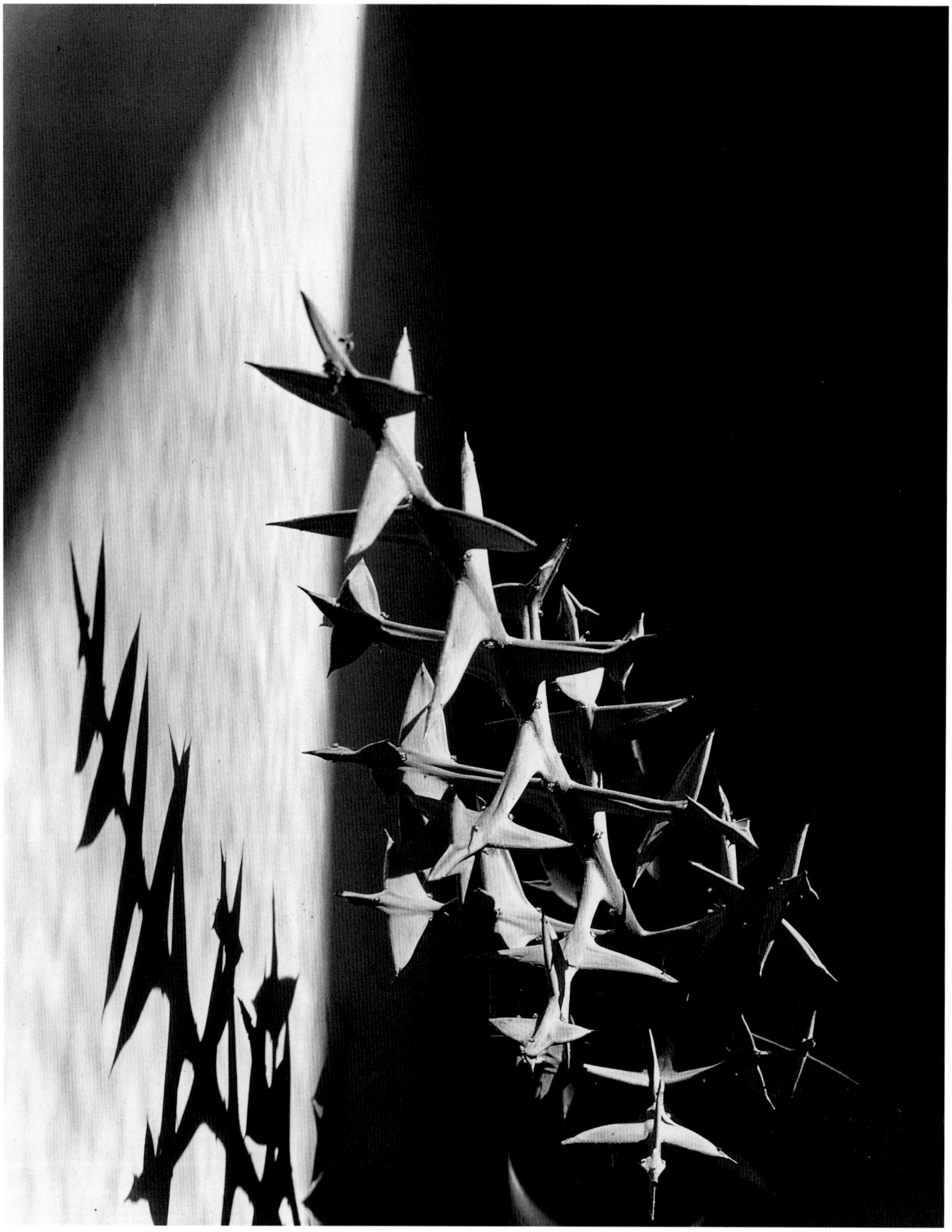

PLATE 36

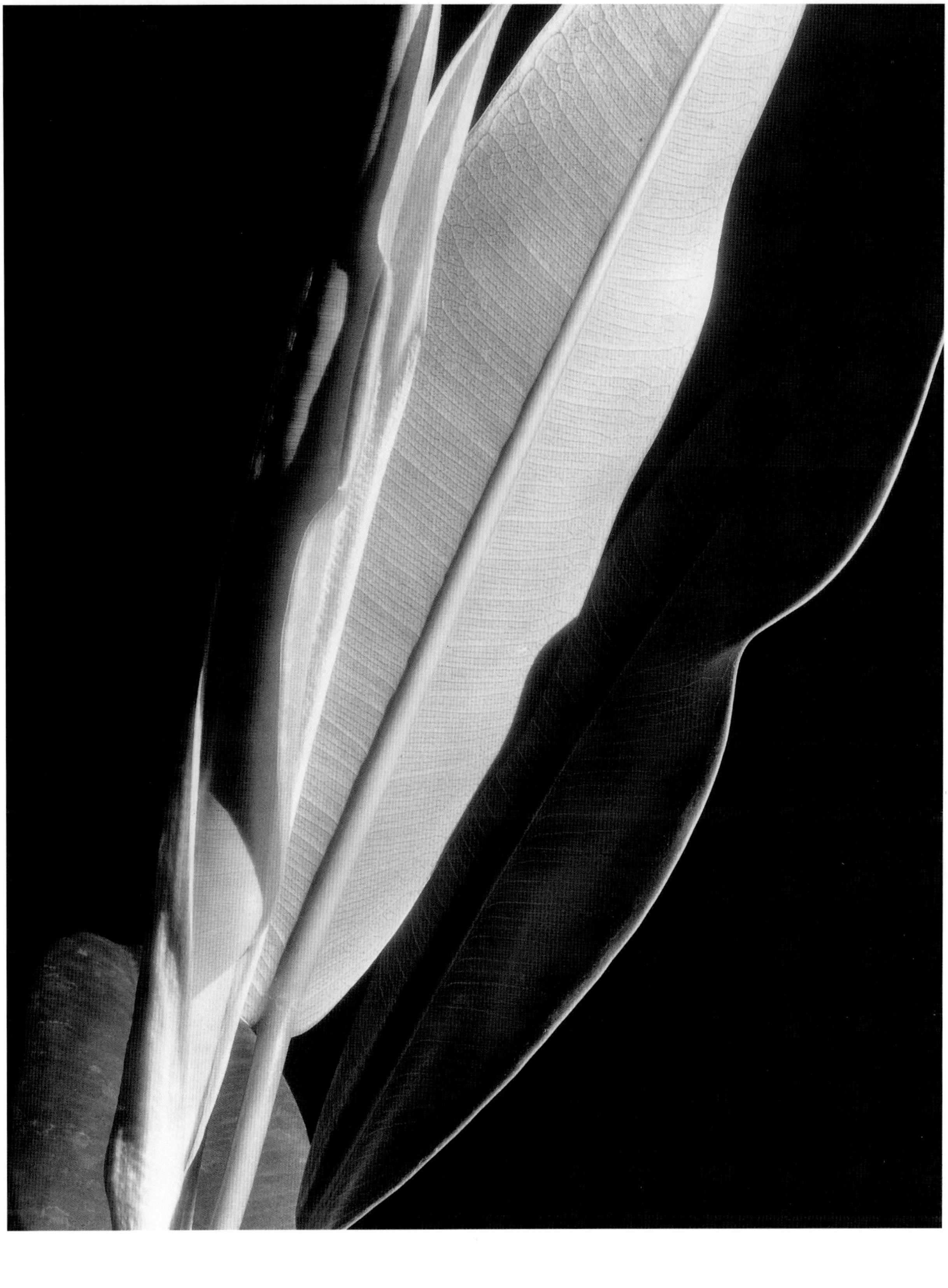

PLATE 37

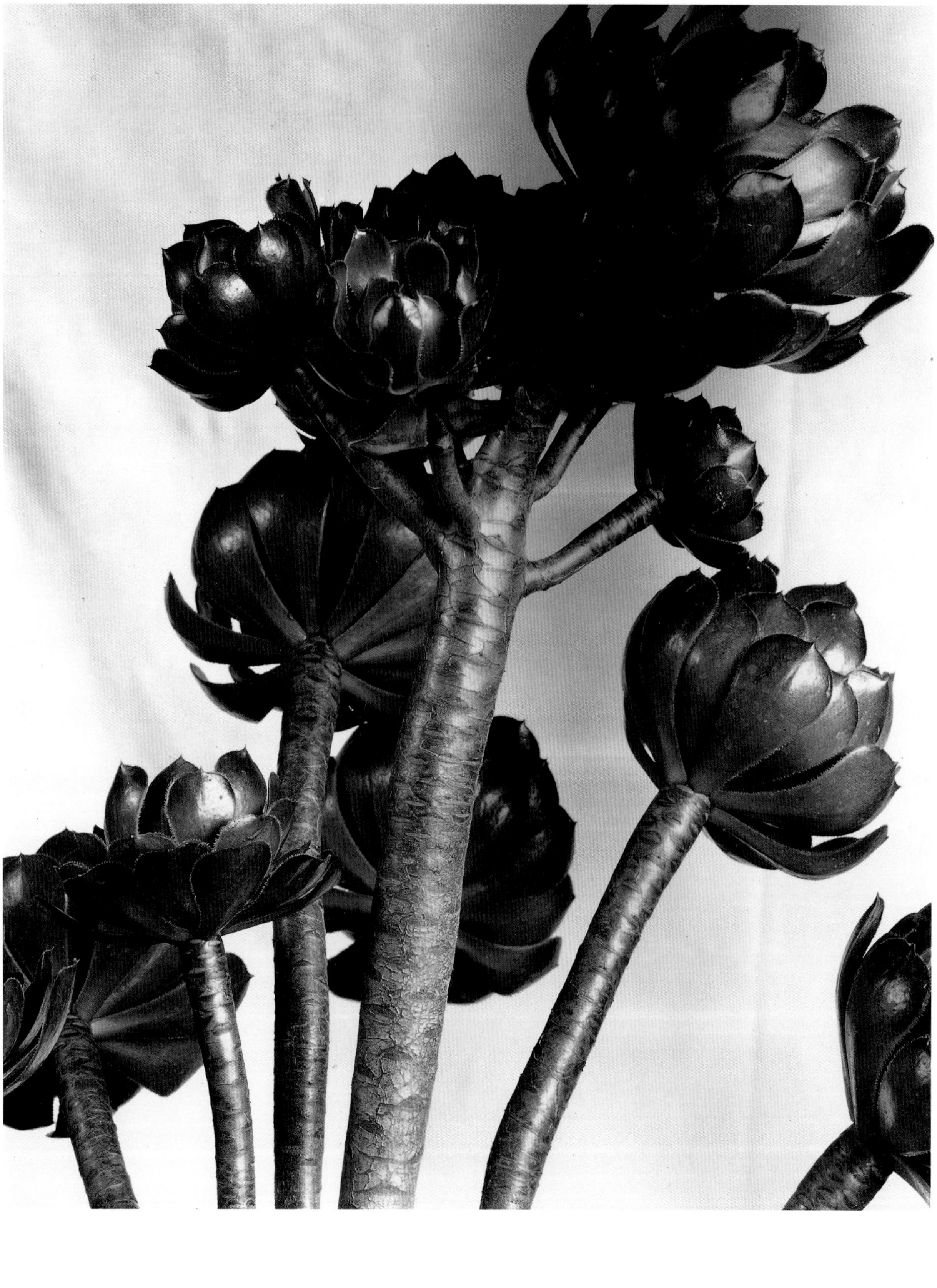

PLATE 38

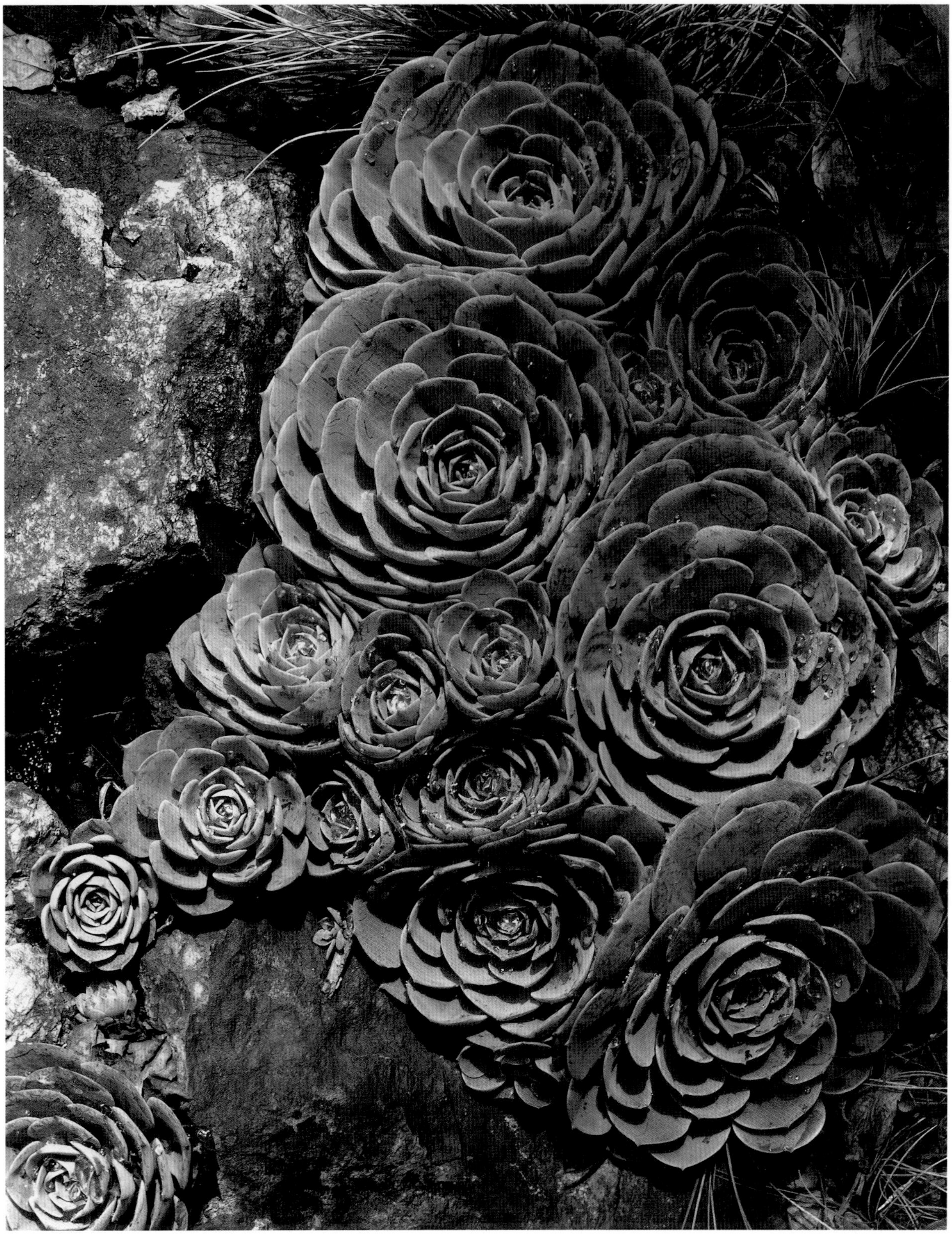

PLATE 39

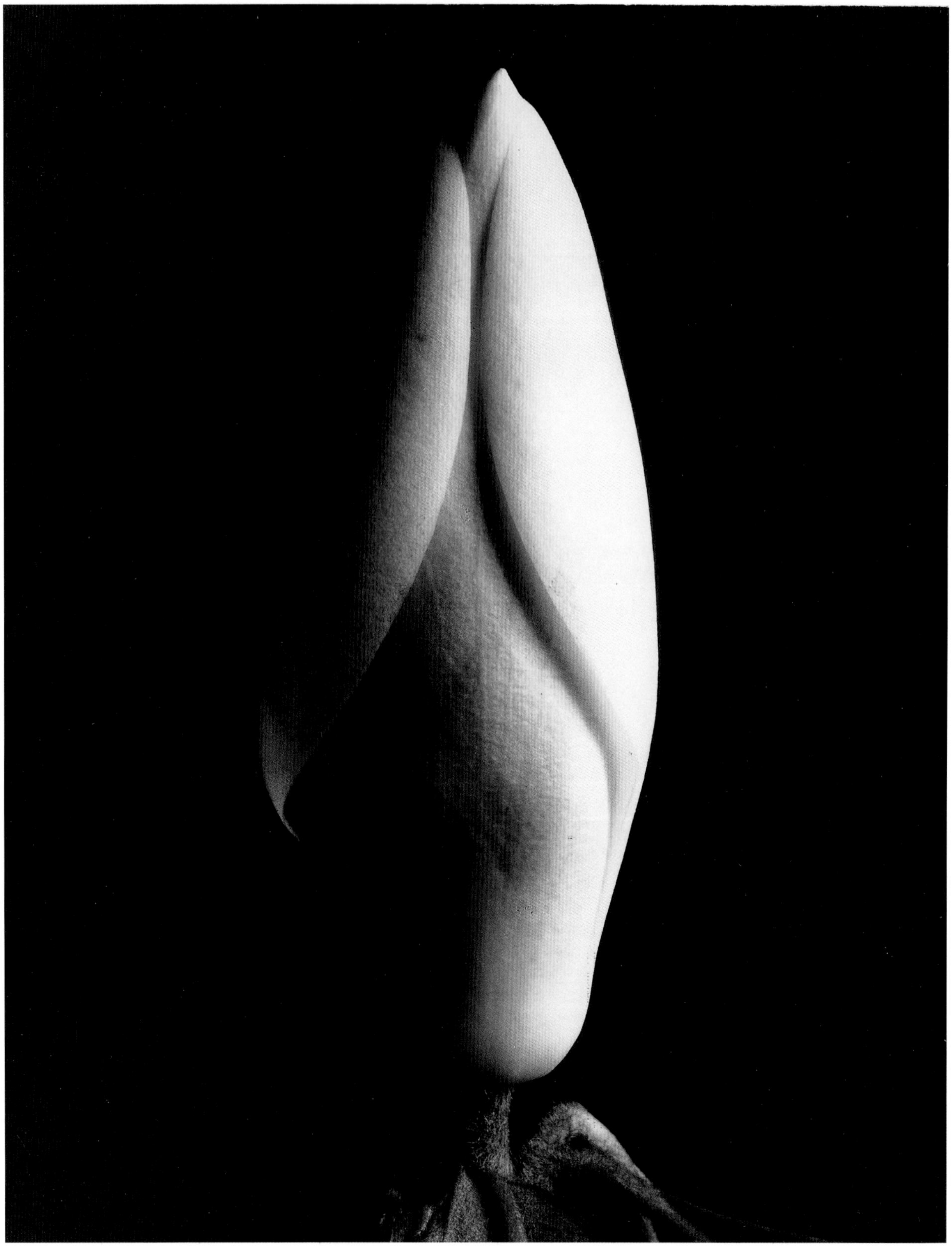

PLATE 40

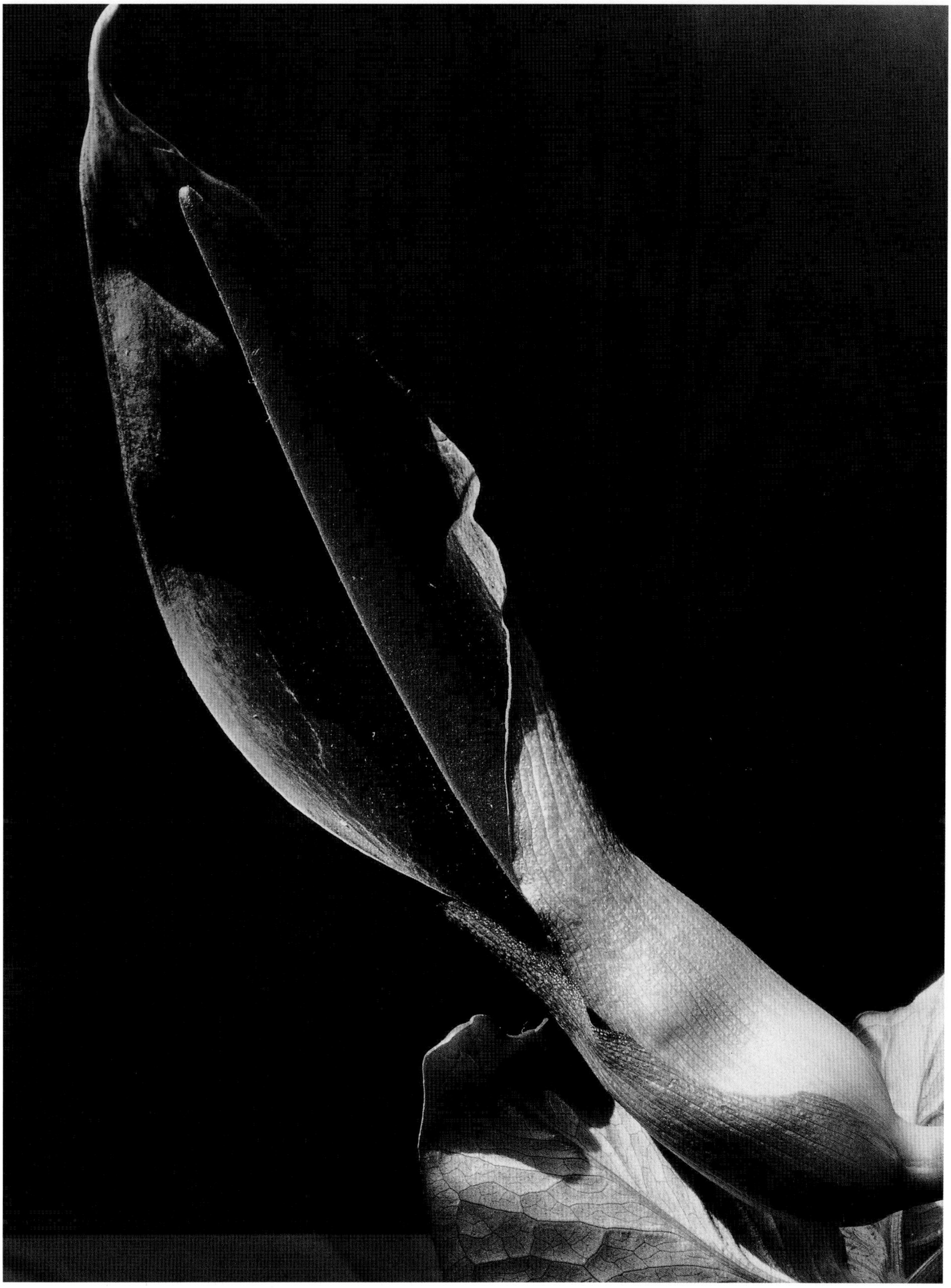

PLATE 41

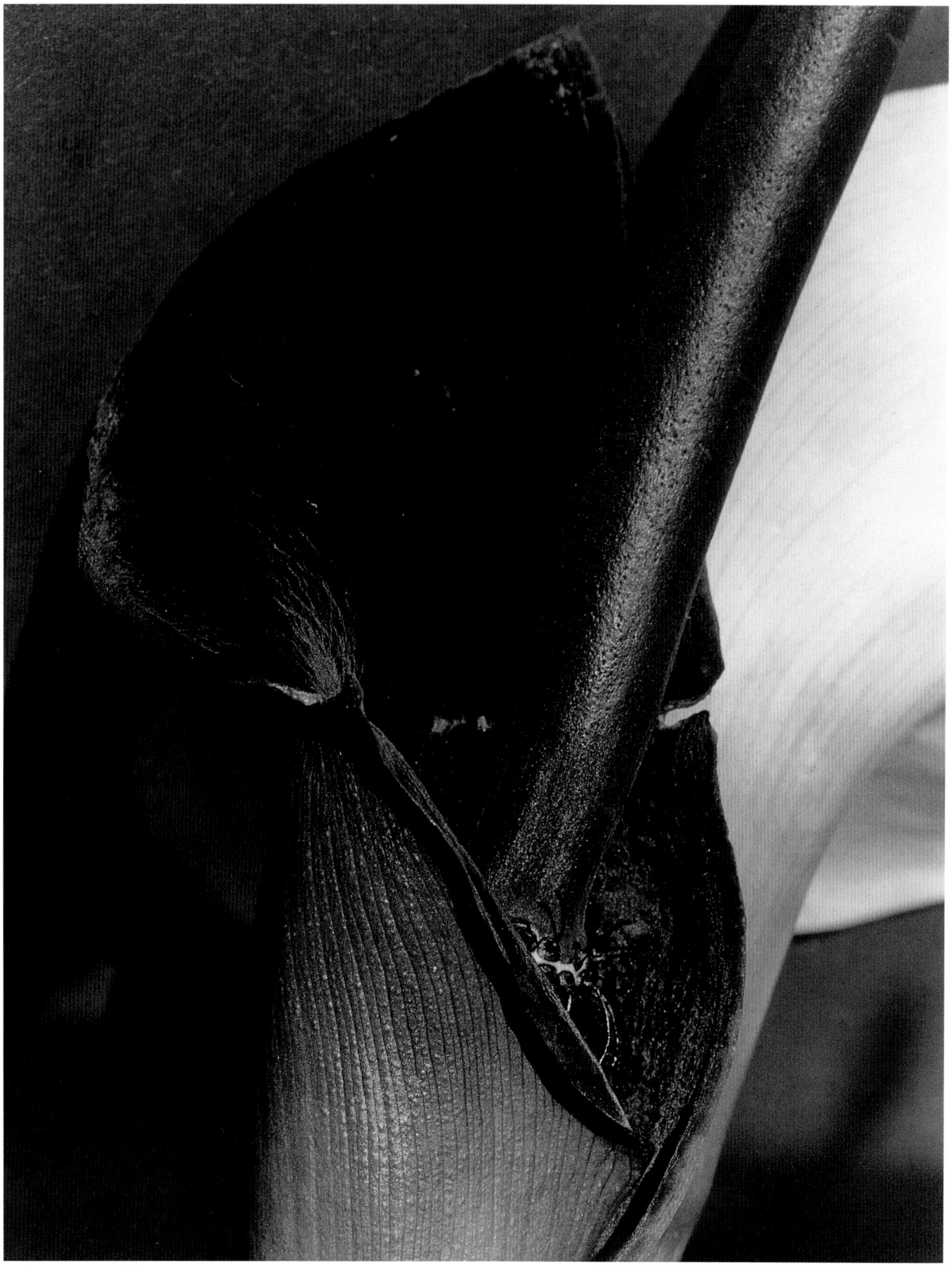

PLATE 42

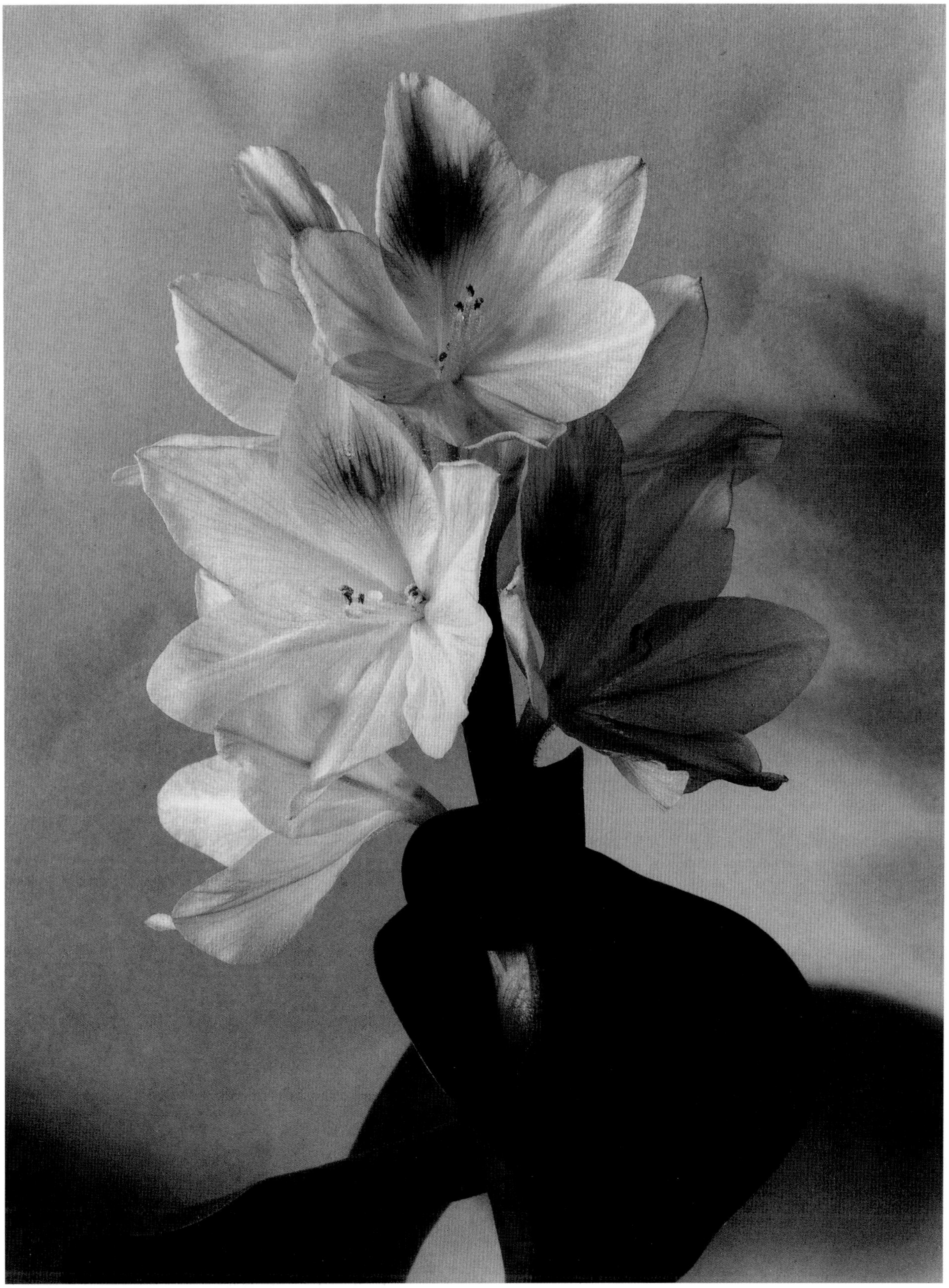

PLATE 43

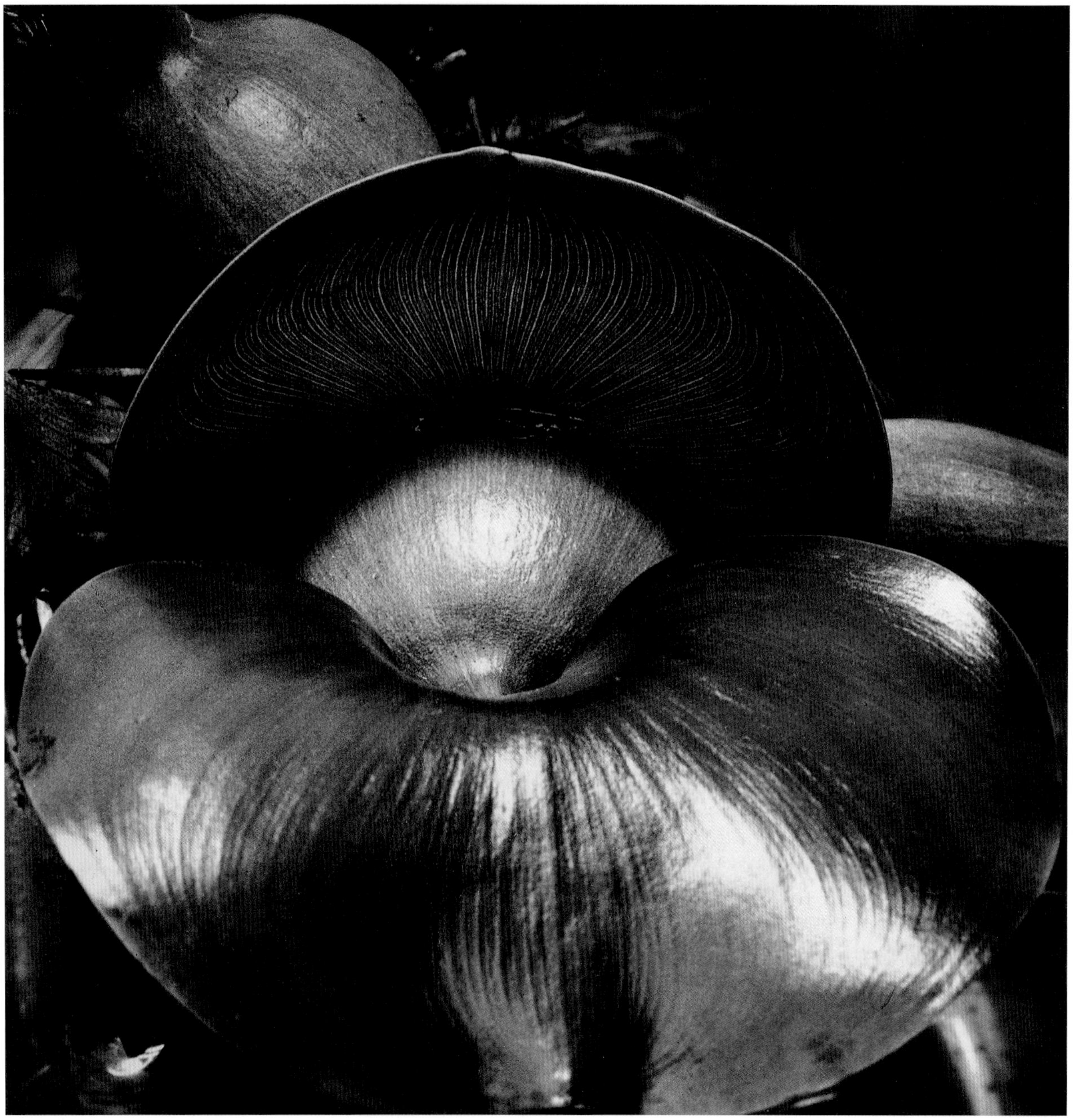

PLATE 44

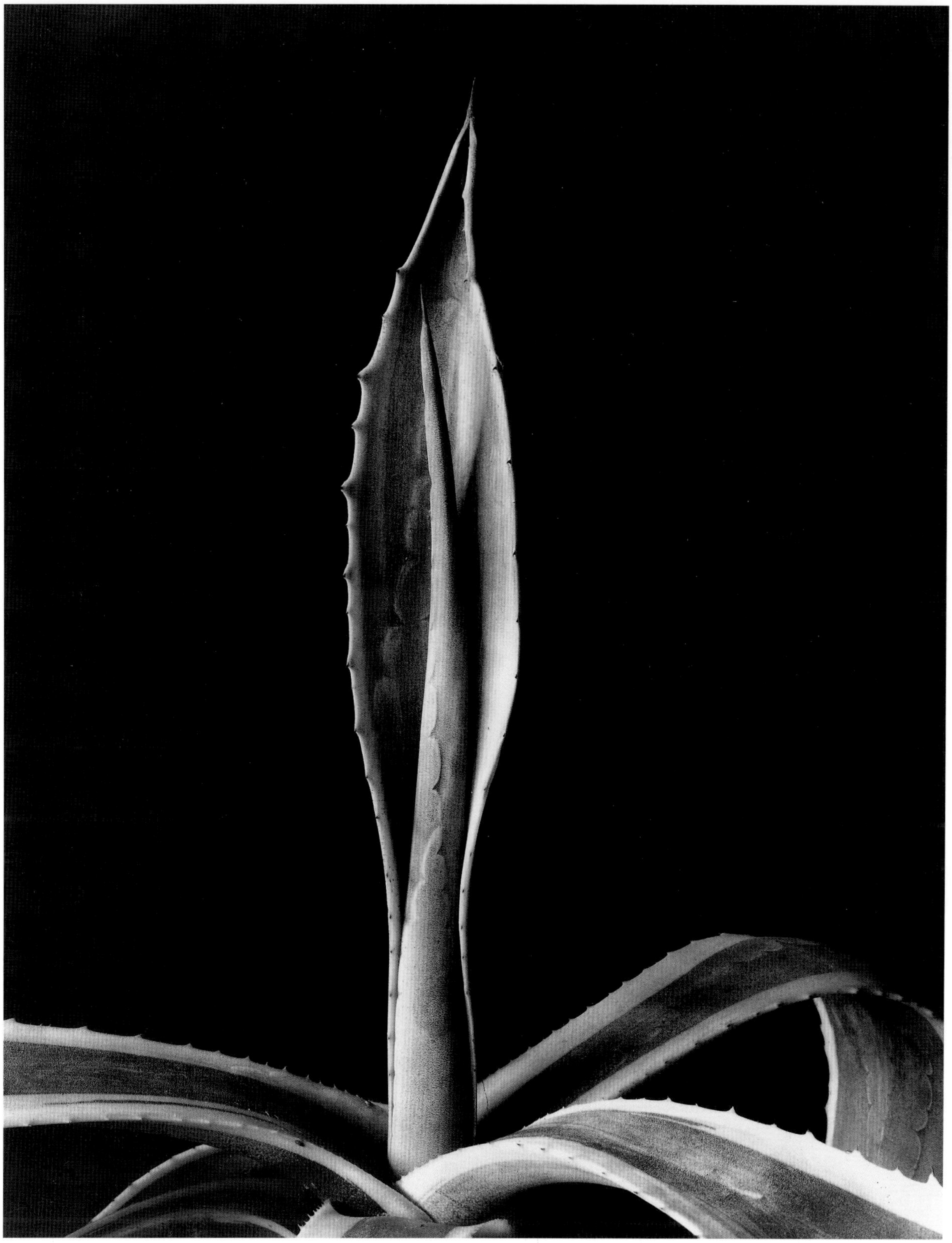

PLATE 45

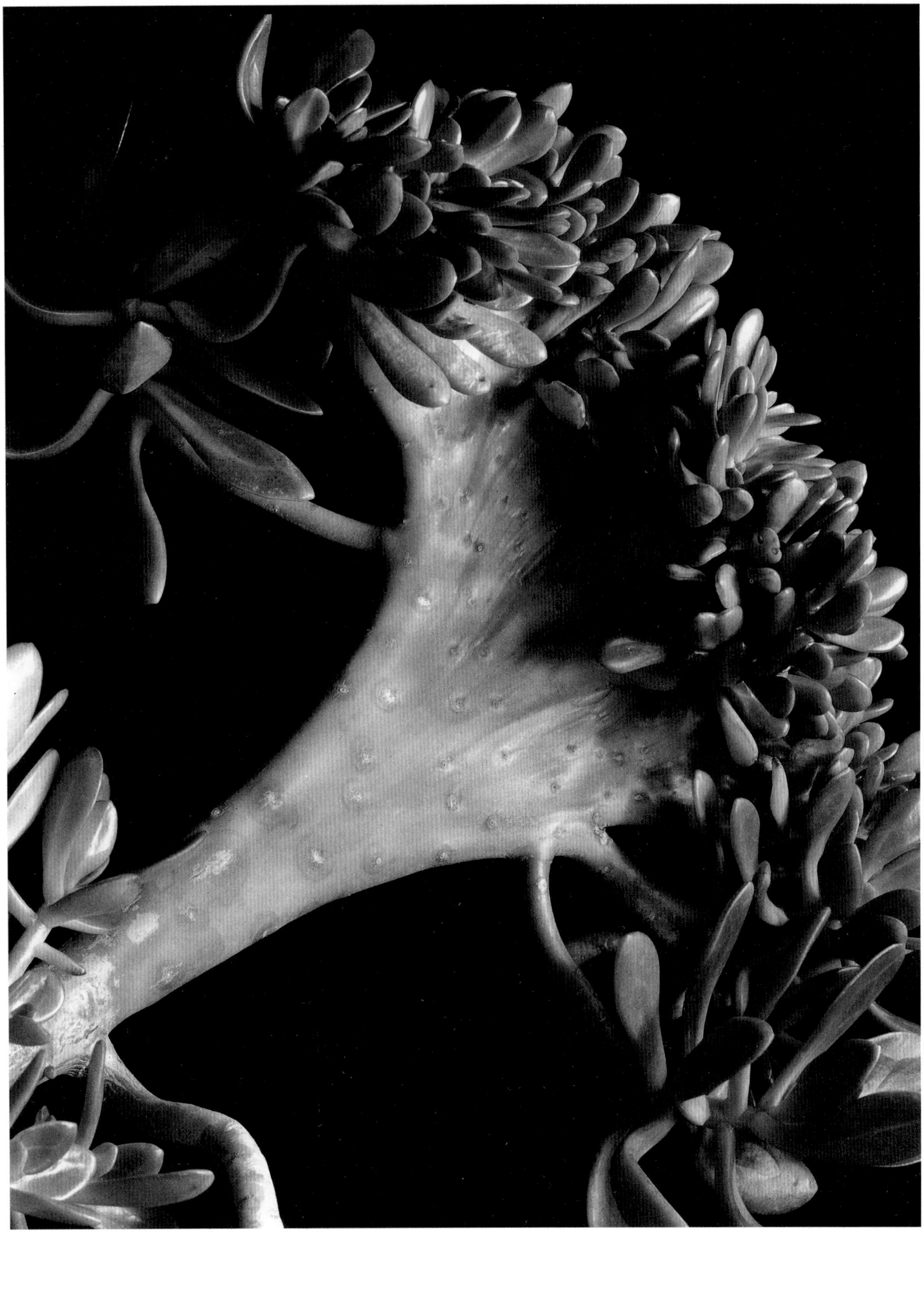

PLATE 46

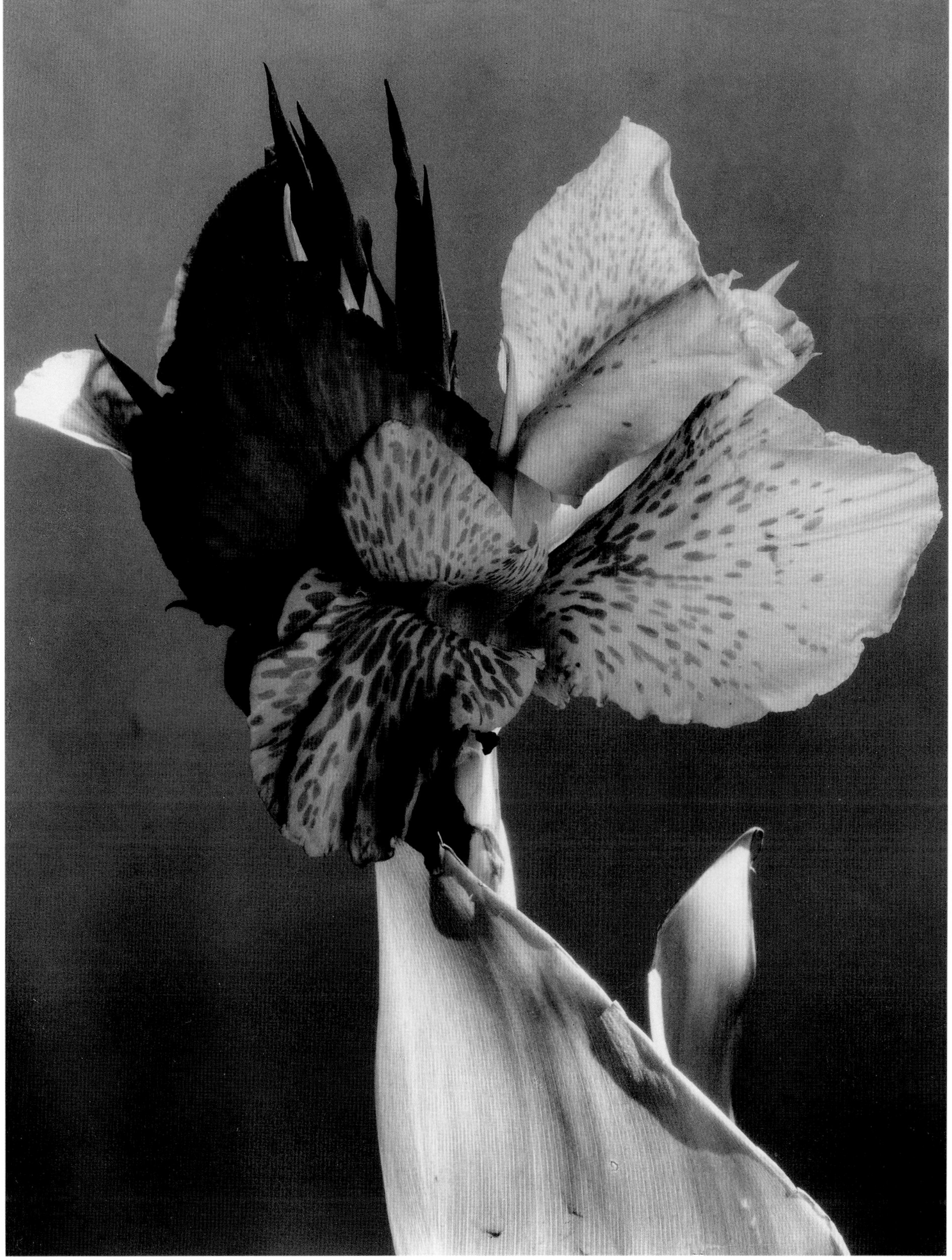

PLATE 47

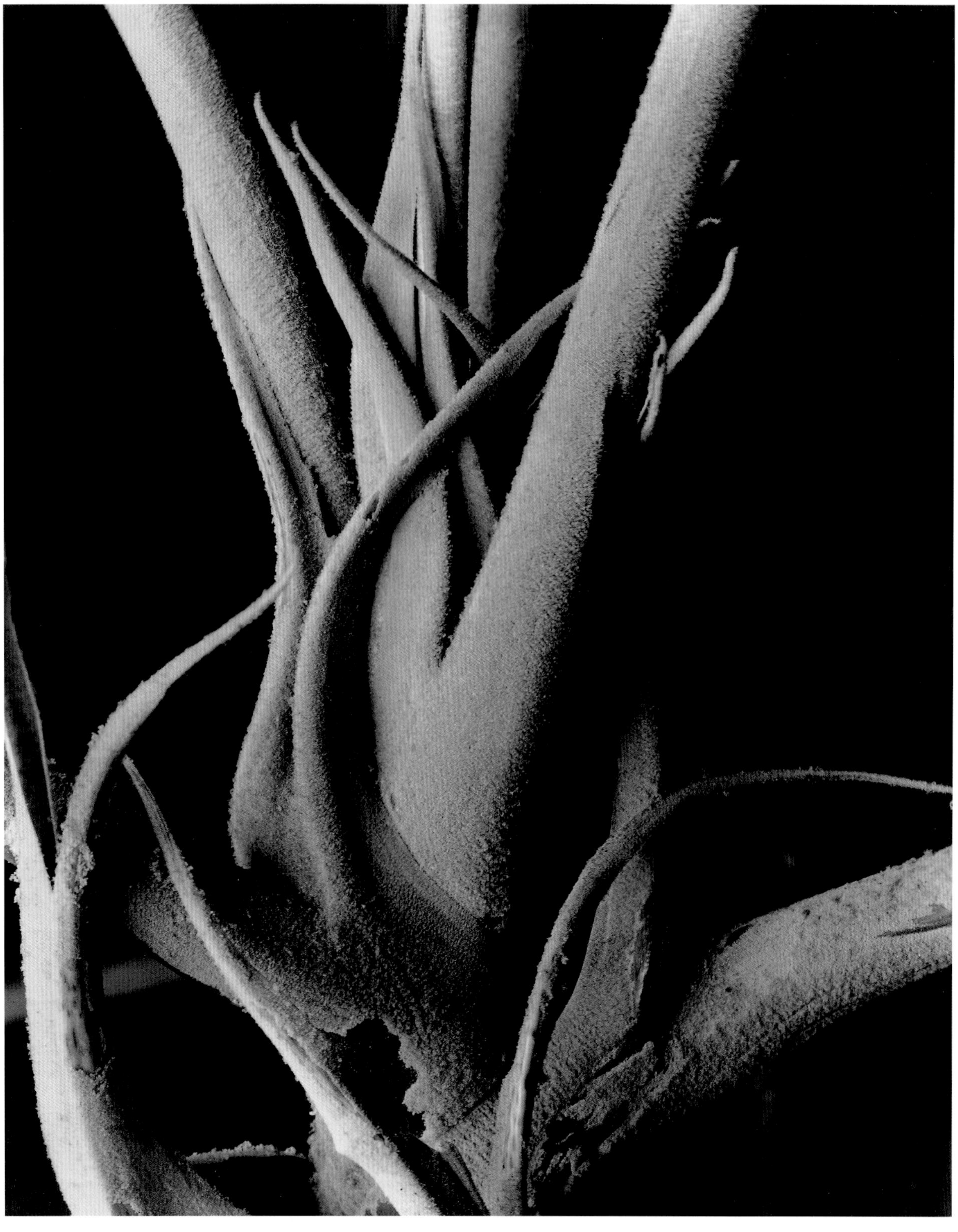

PLATE 48

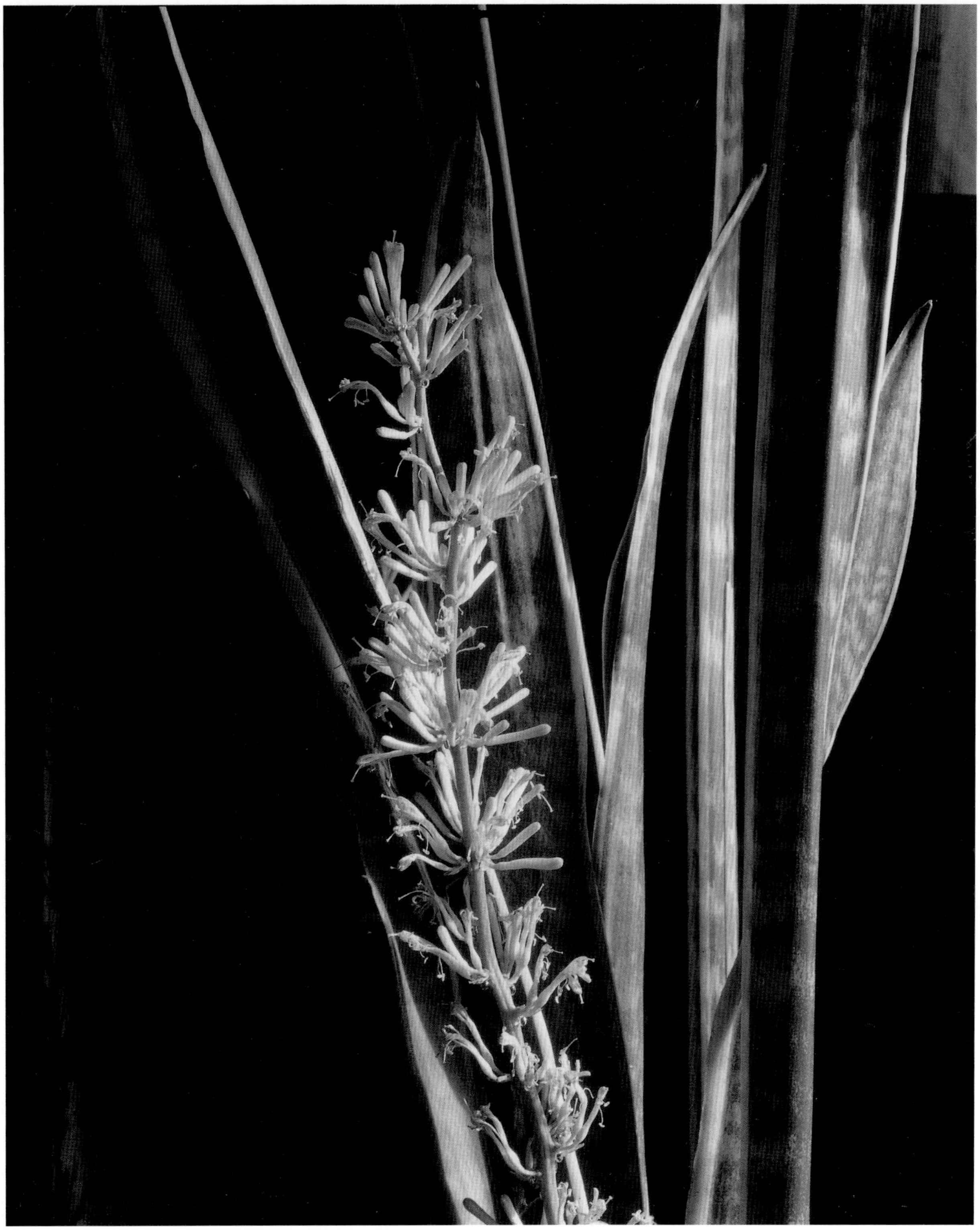

PLATE 49

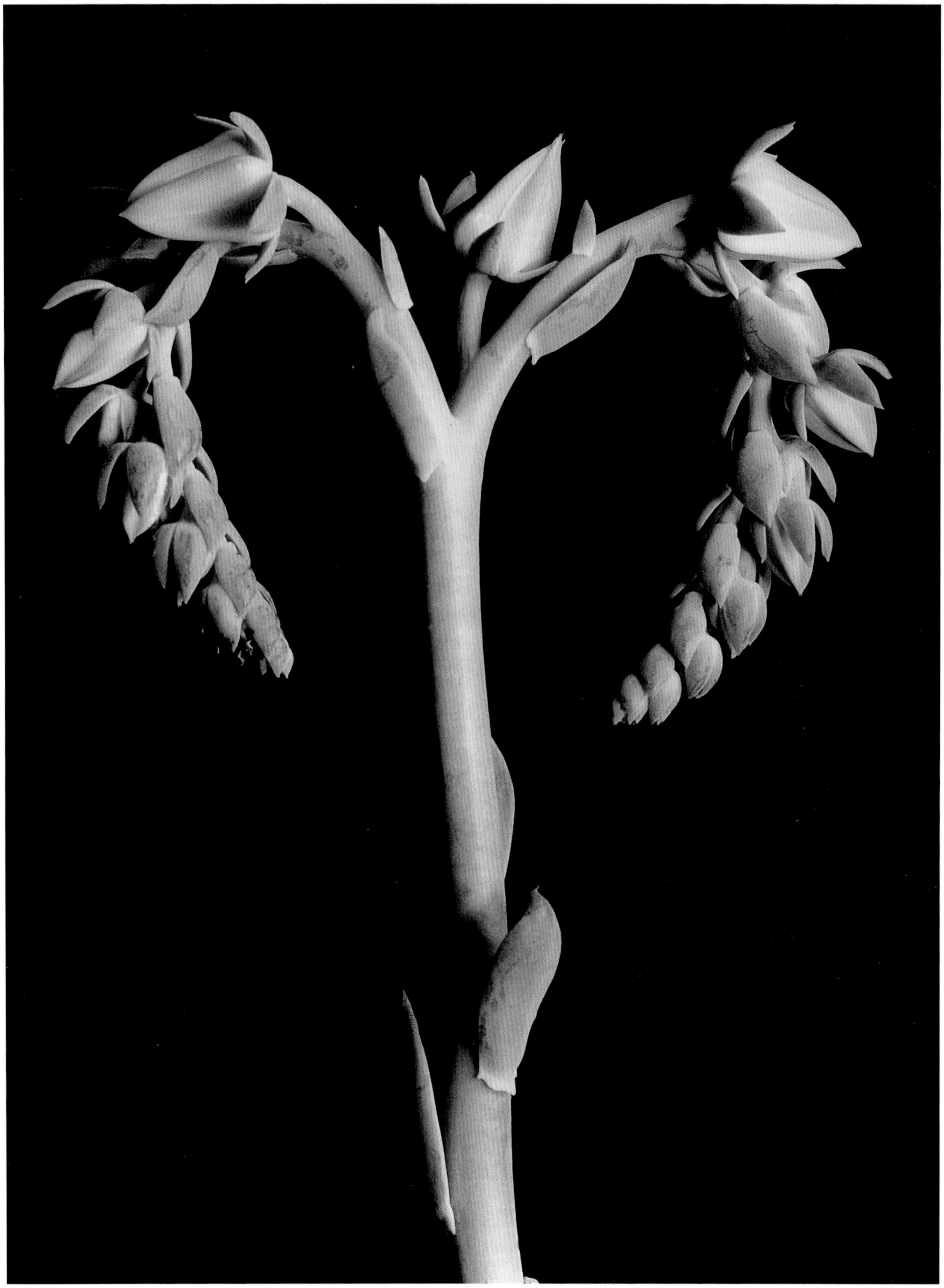

PLATE 50

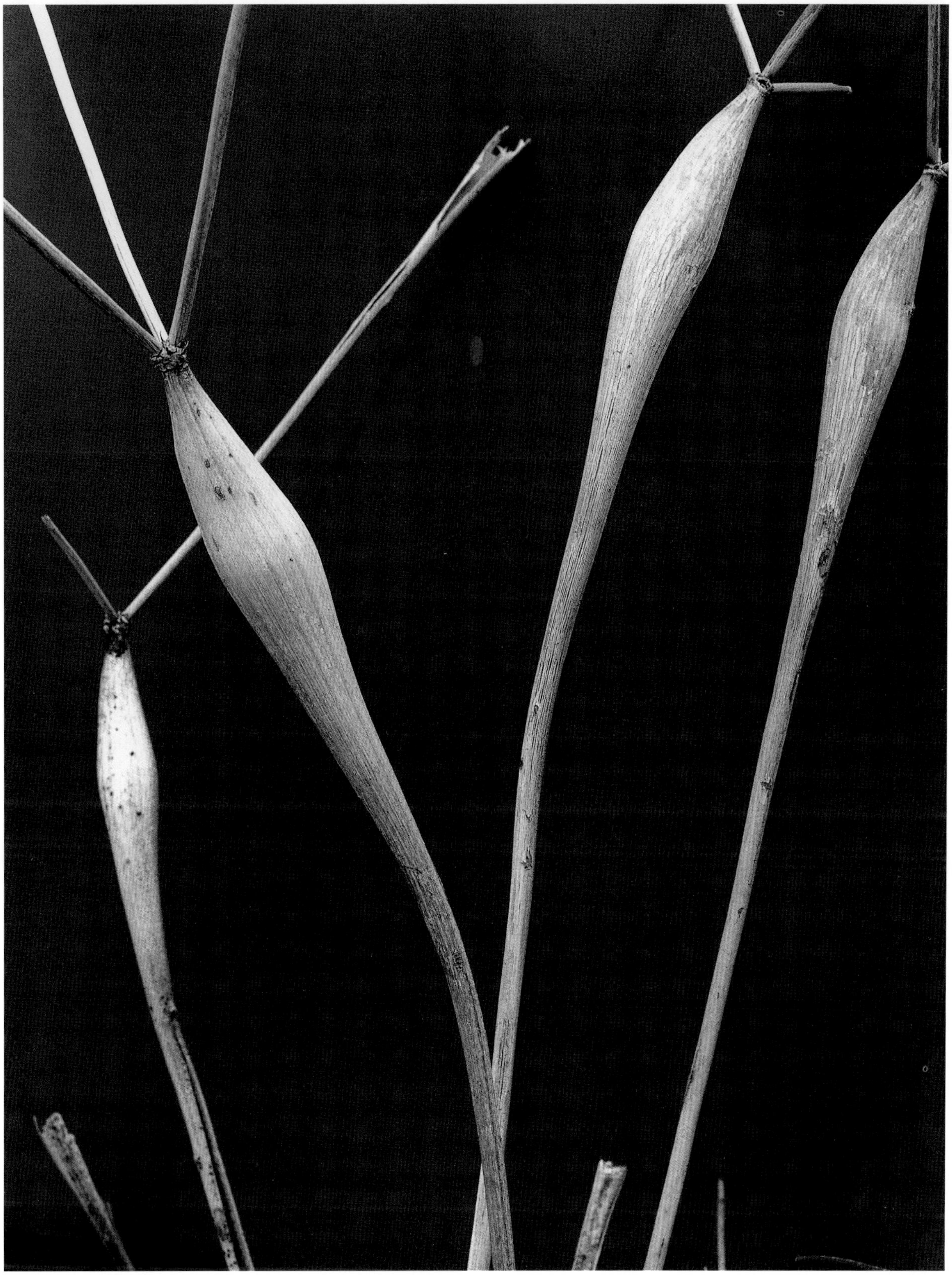

PLATE 51

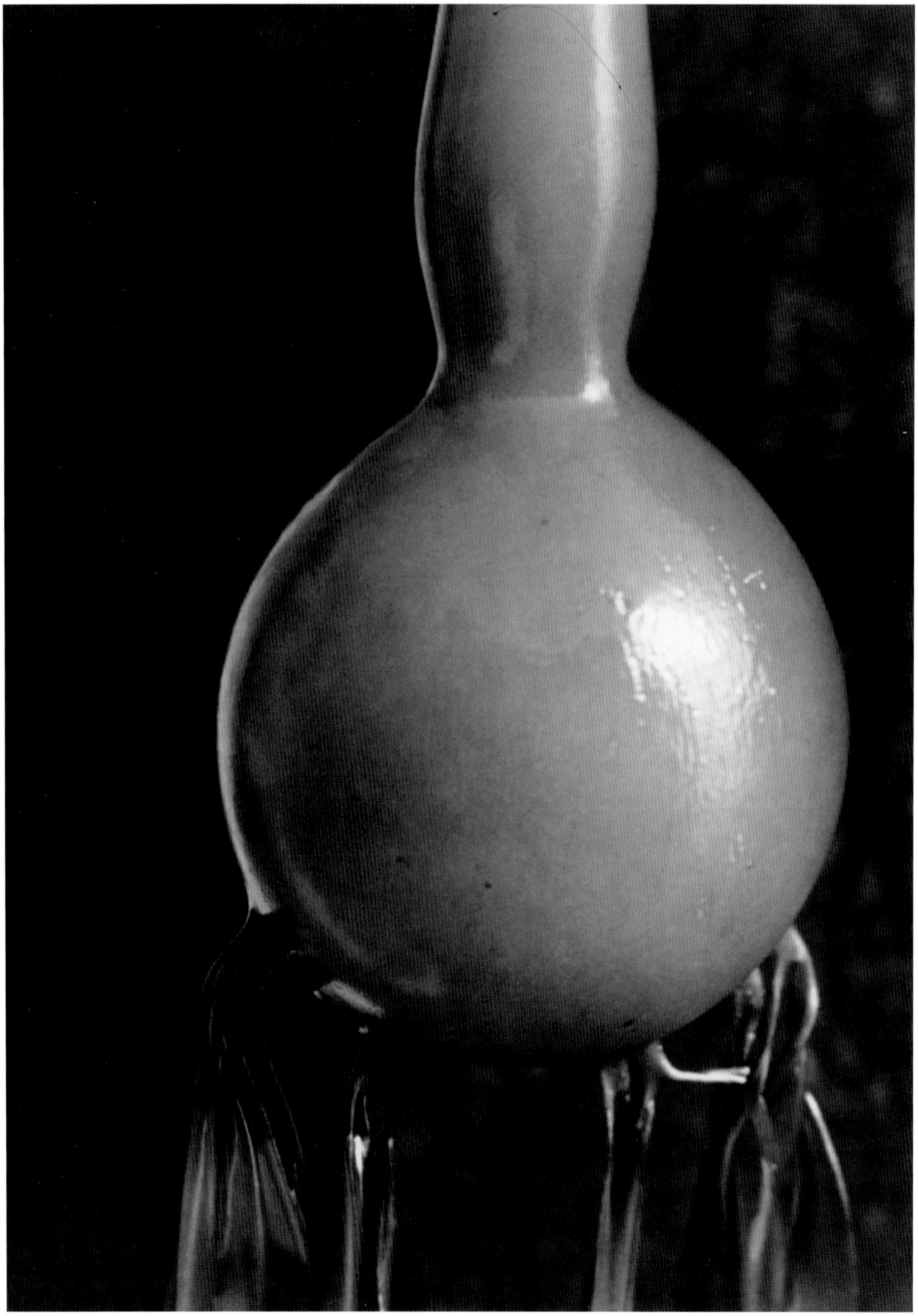

PLATE 52

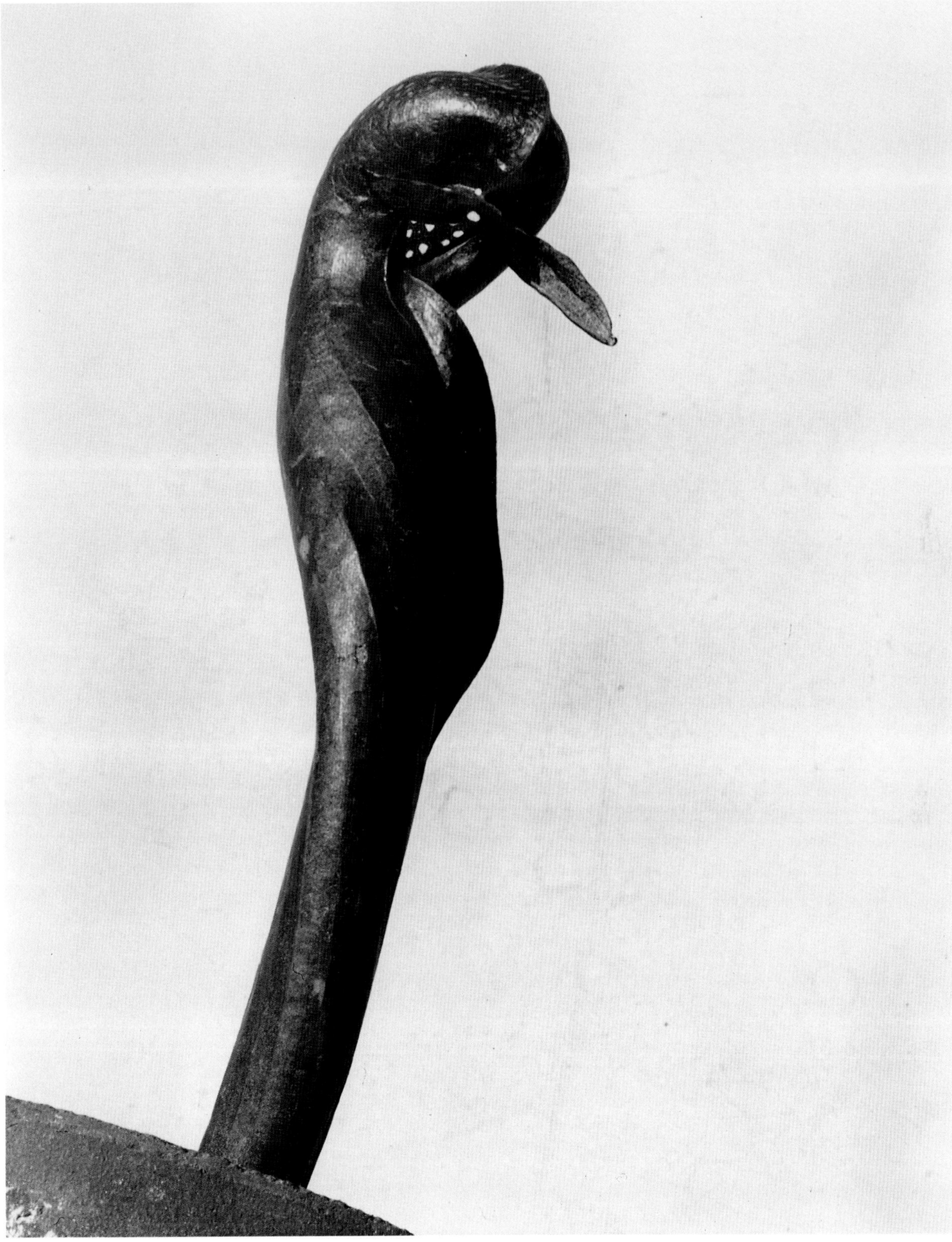

PLATE 53

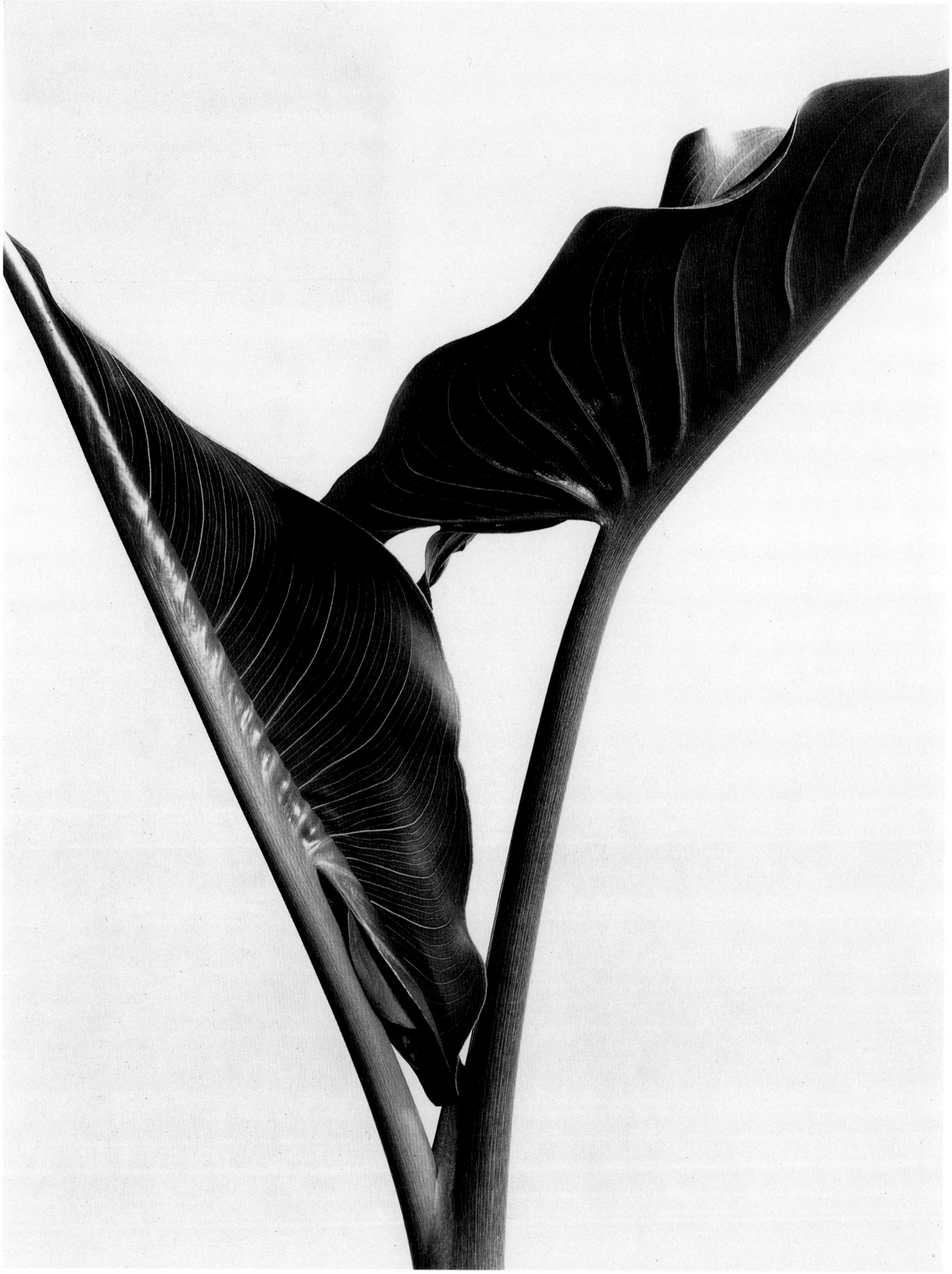

PLATE 54

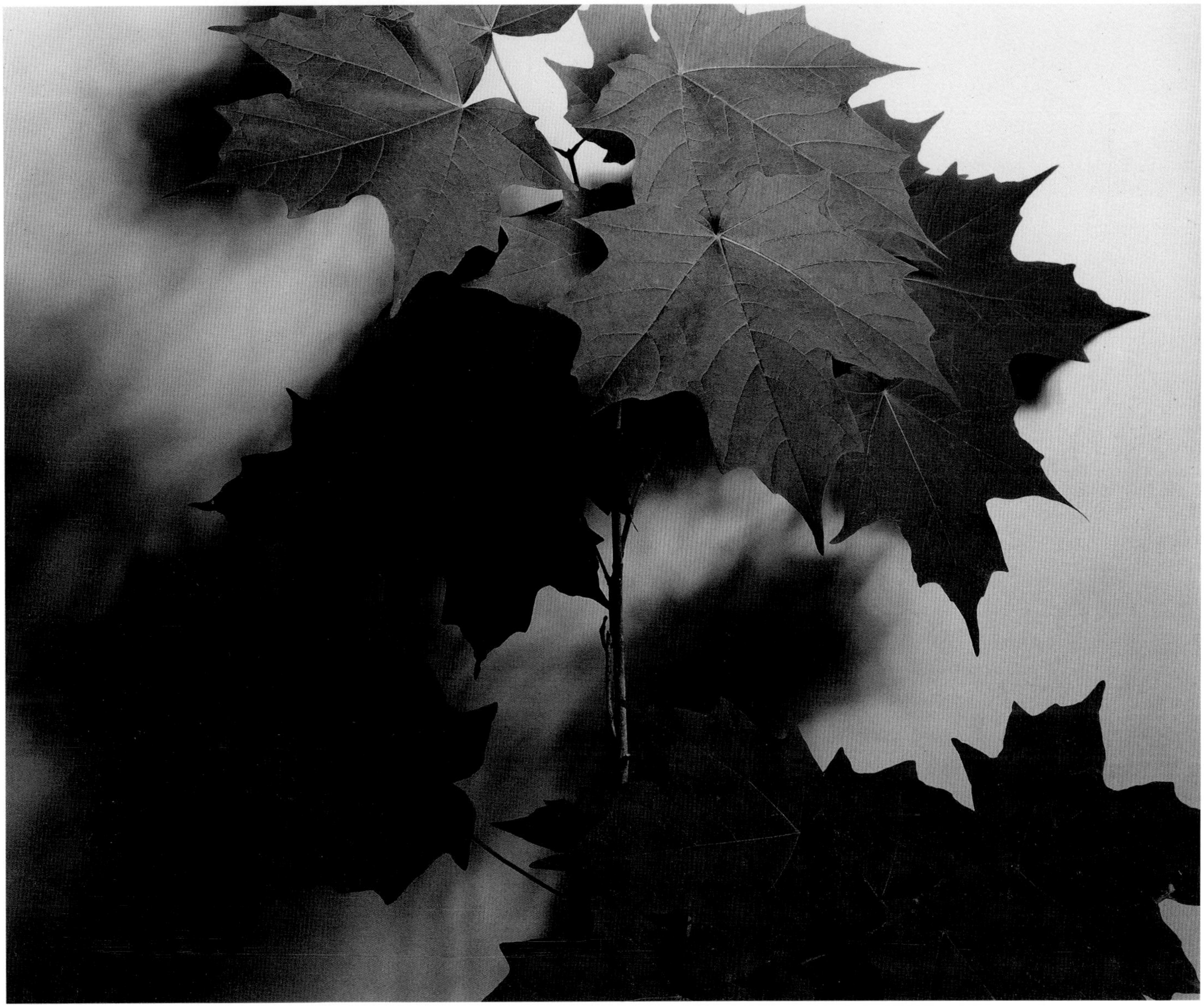

PLATE 55

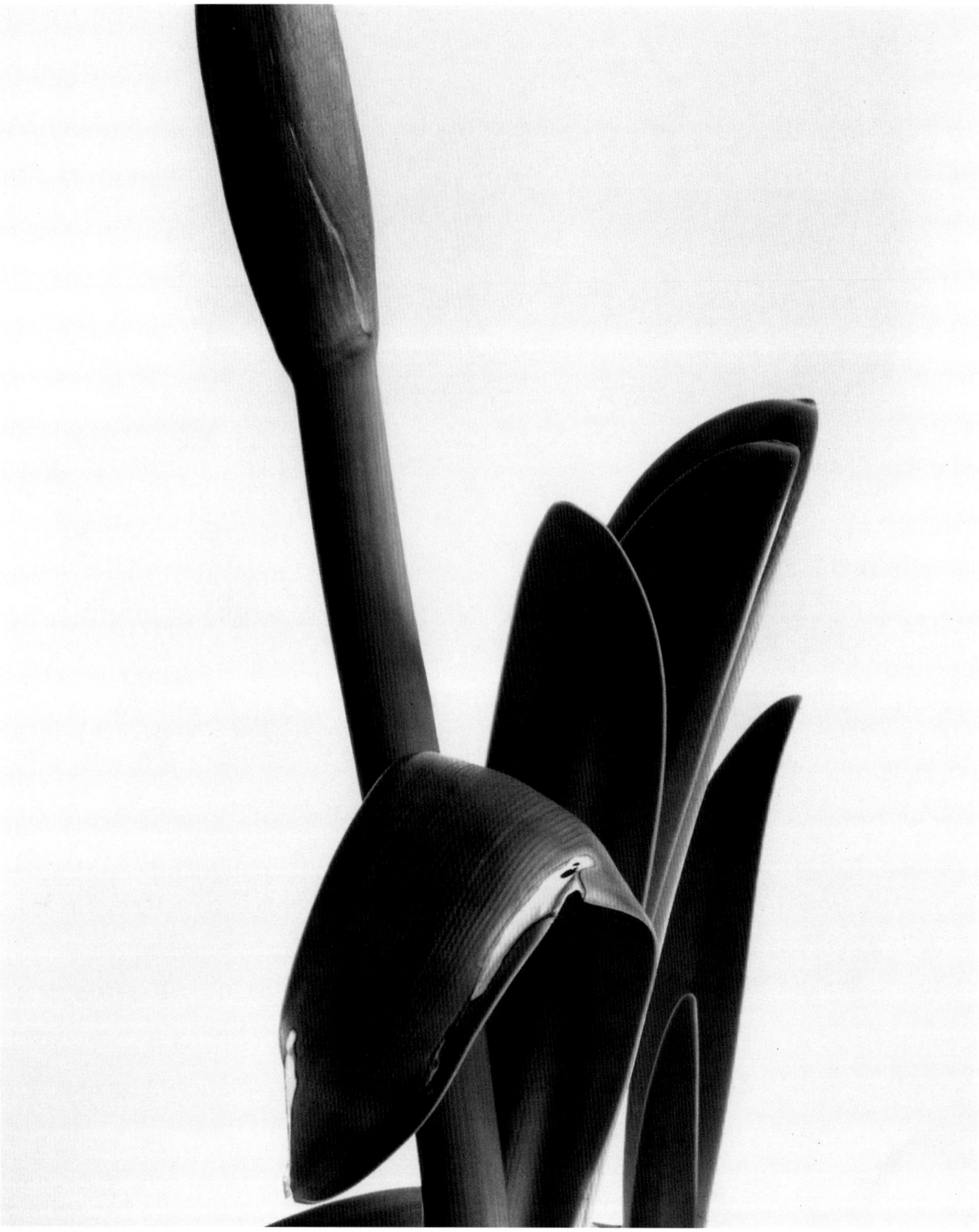

PLATE 56

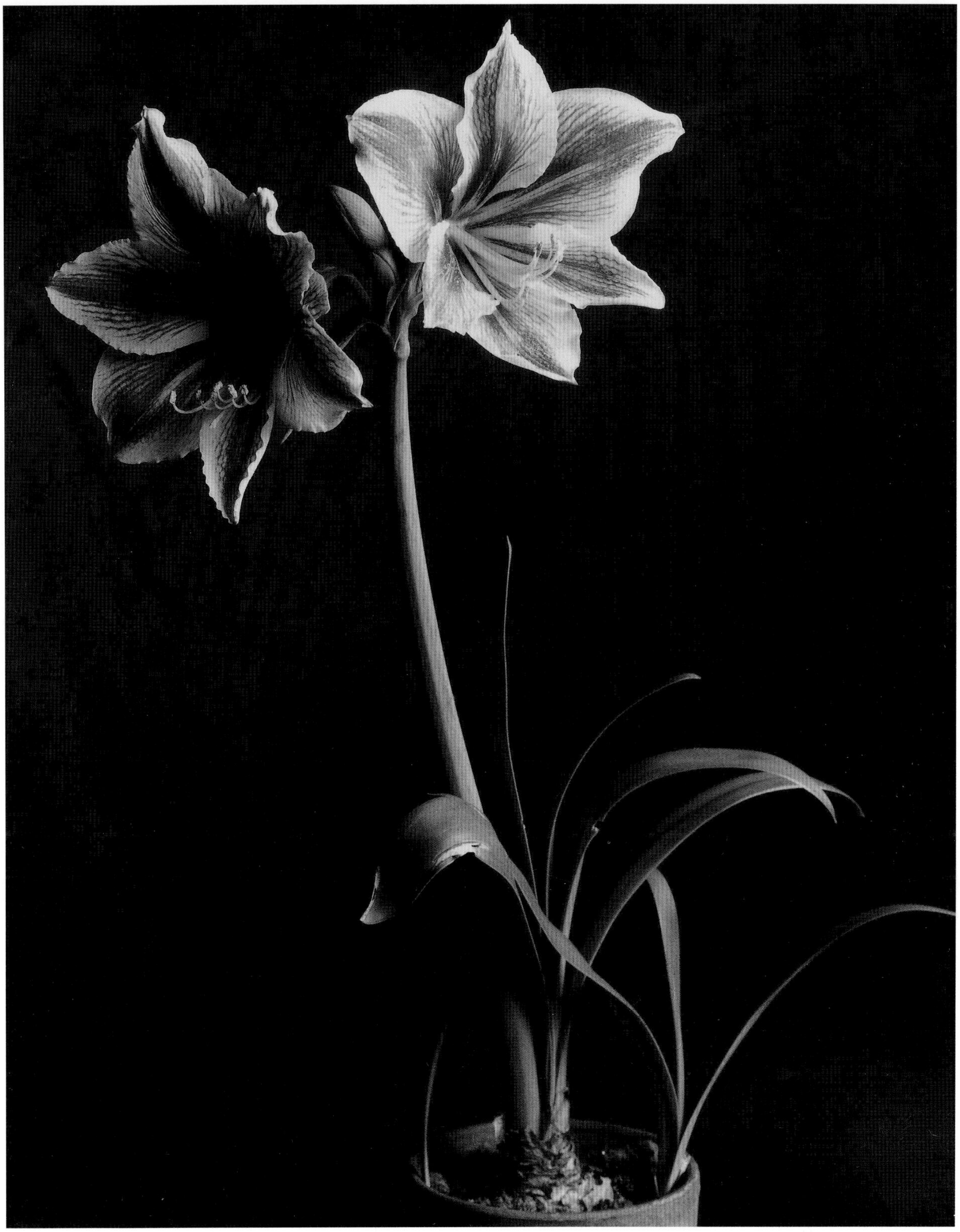

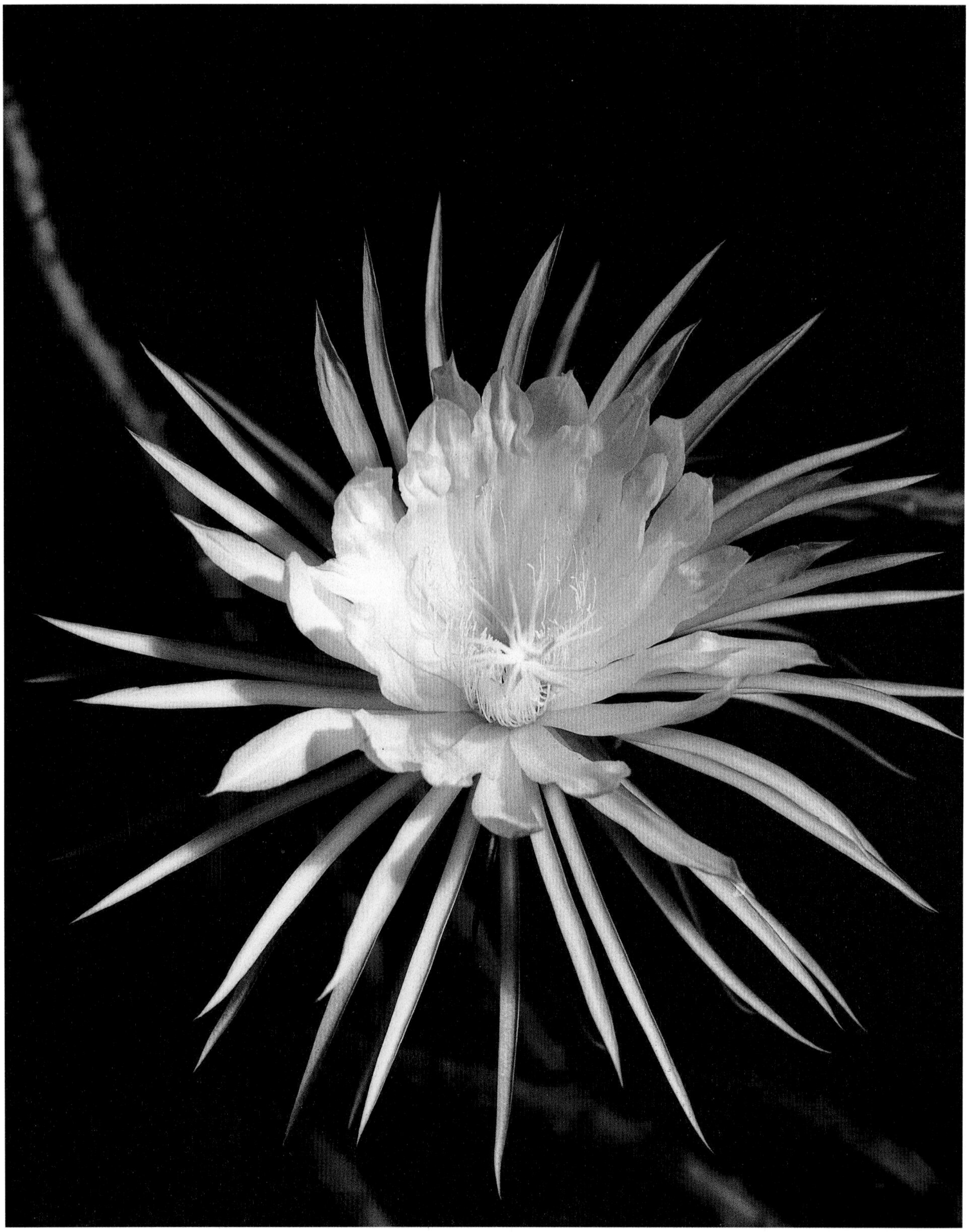

PLATE 58

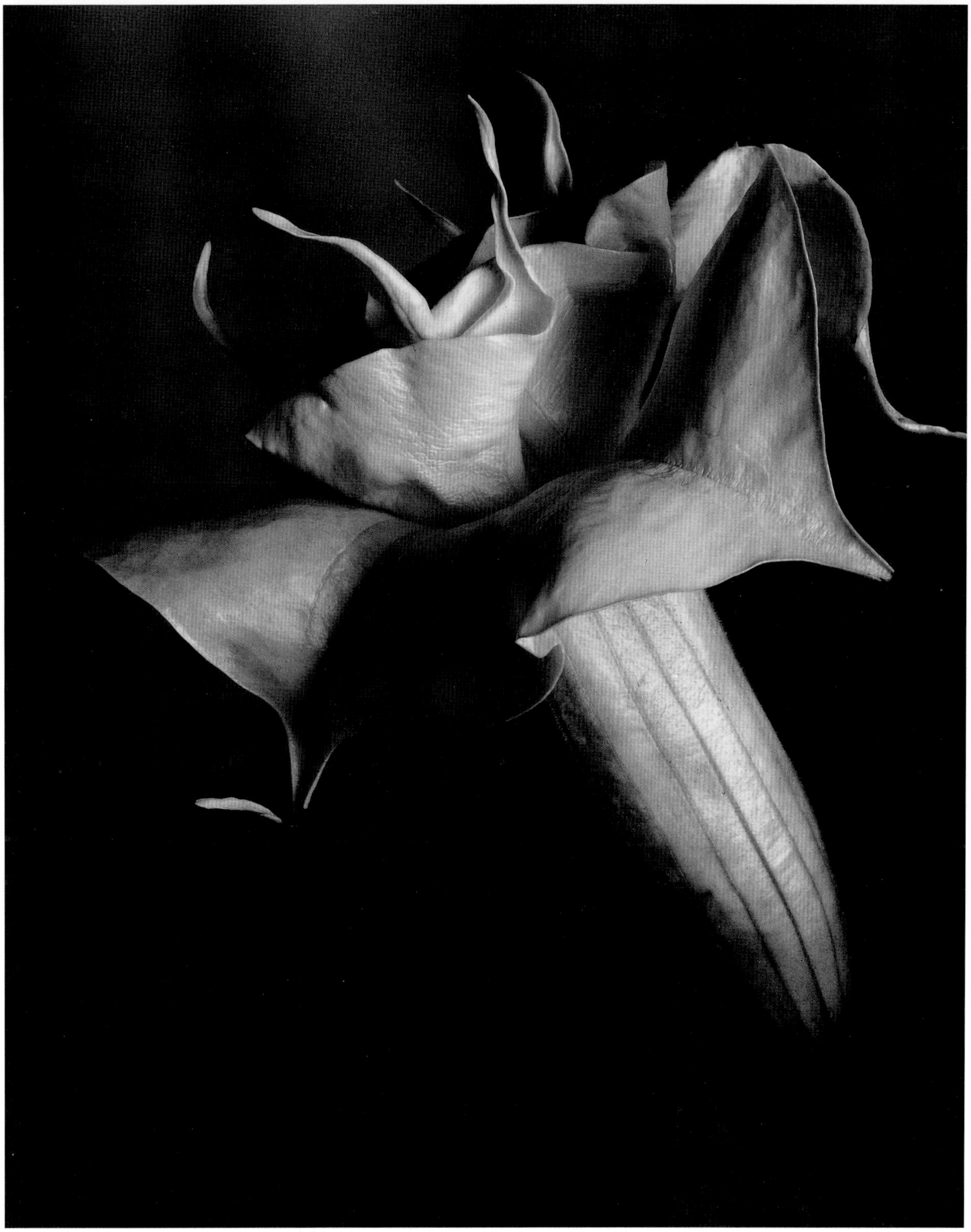

PLATE 59

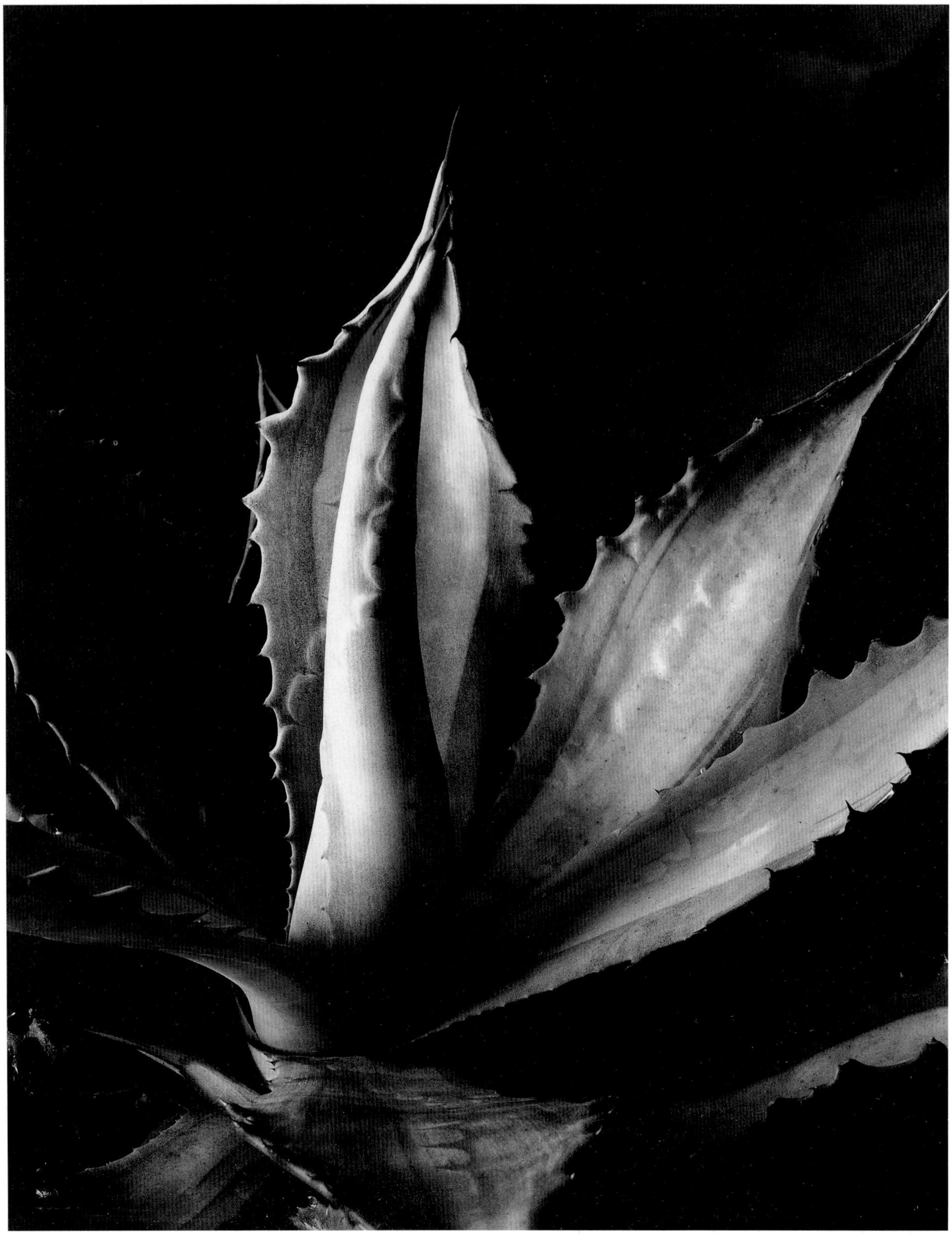

PLATE 60

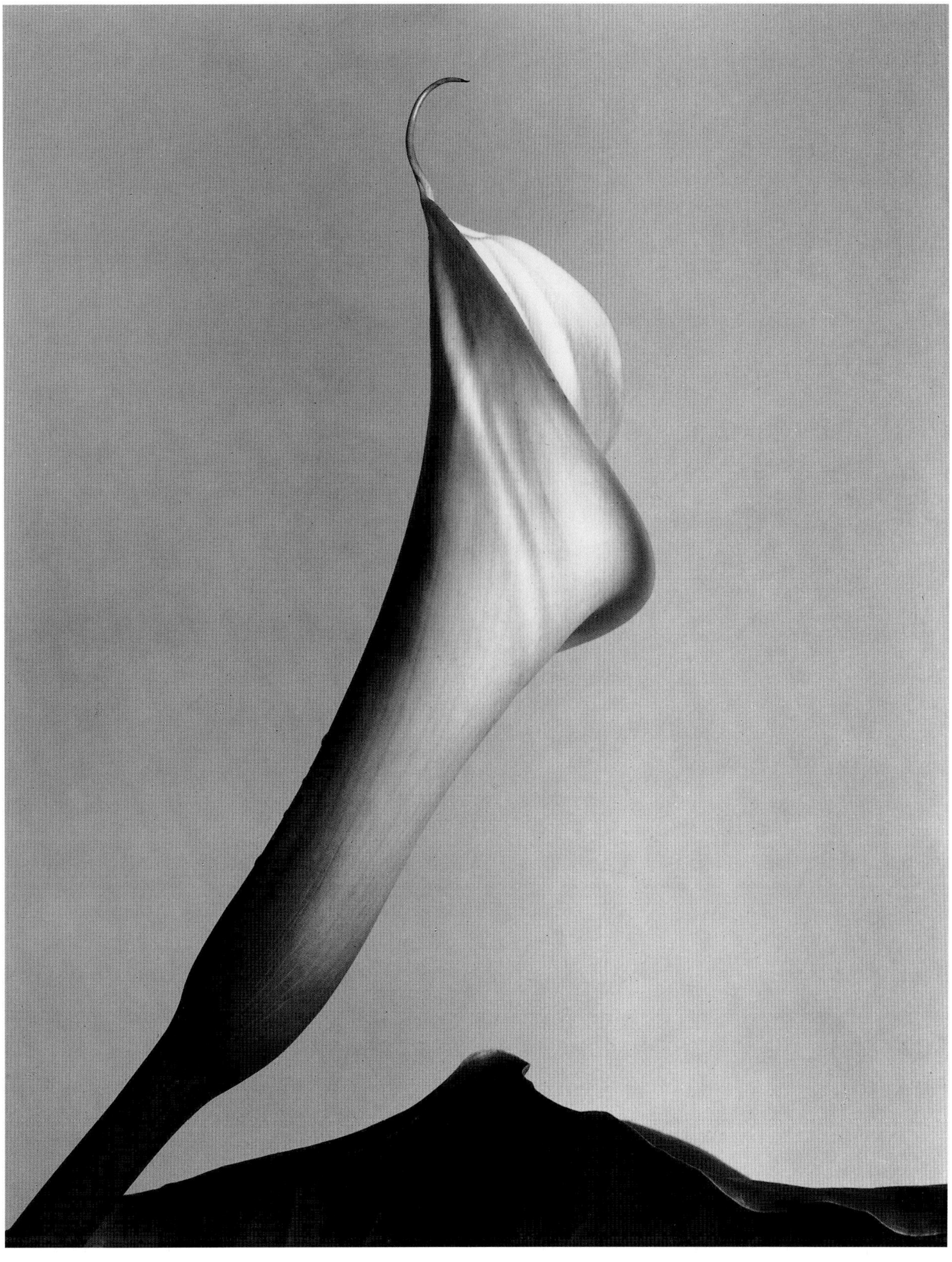

PLATE 61

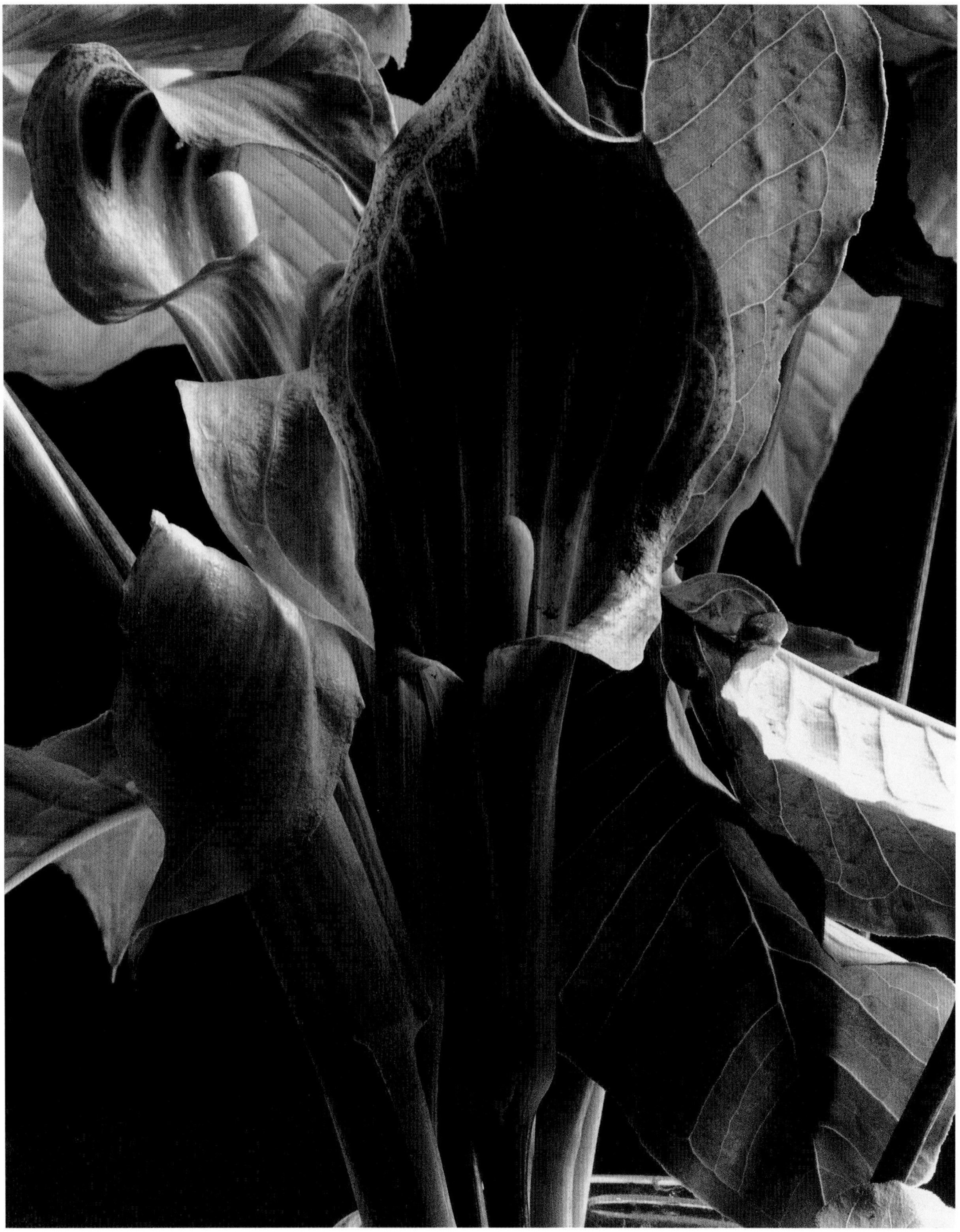

PLATE 62

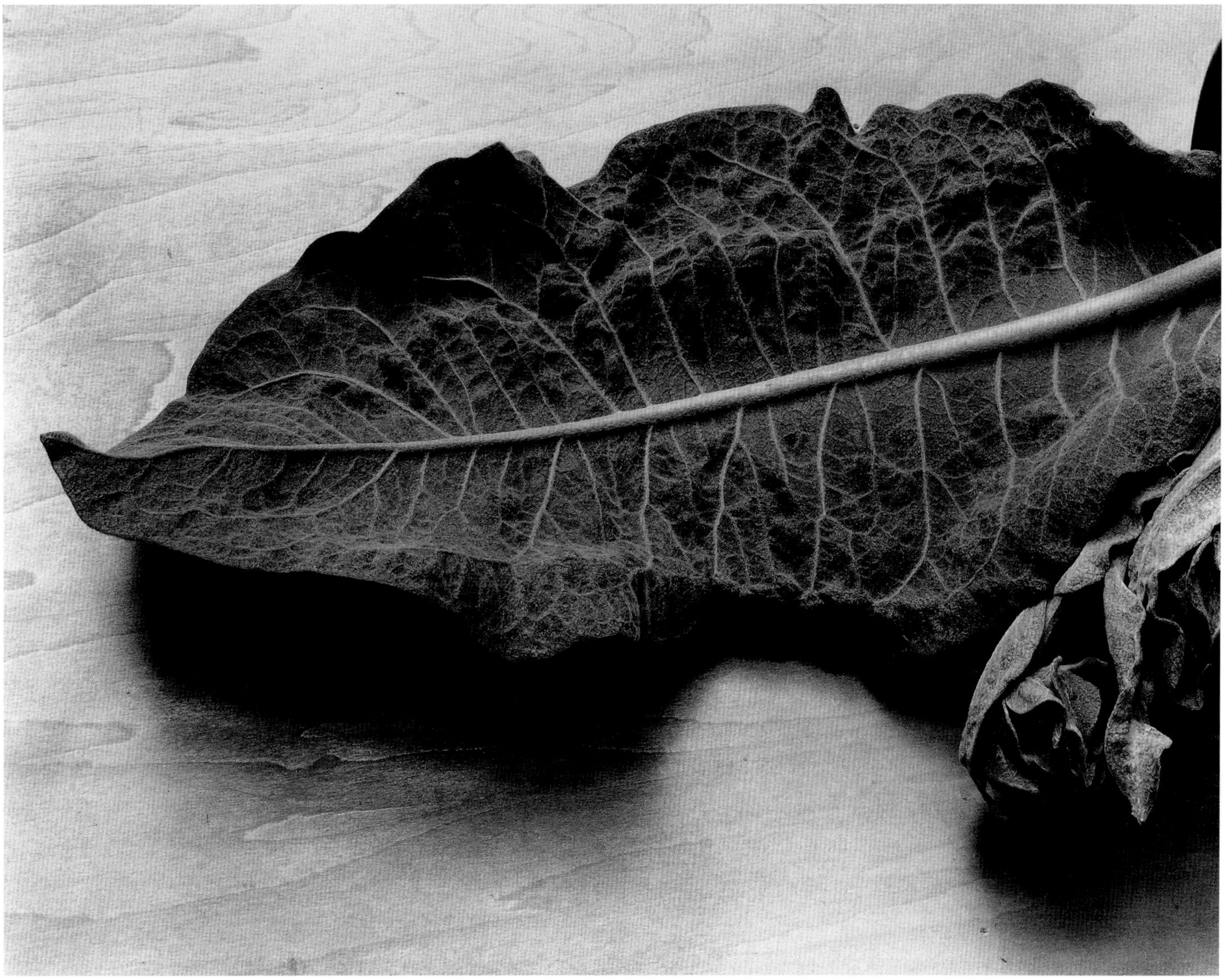

PLATE 63

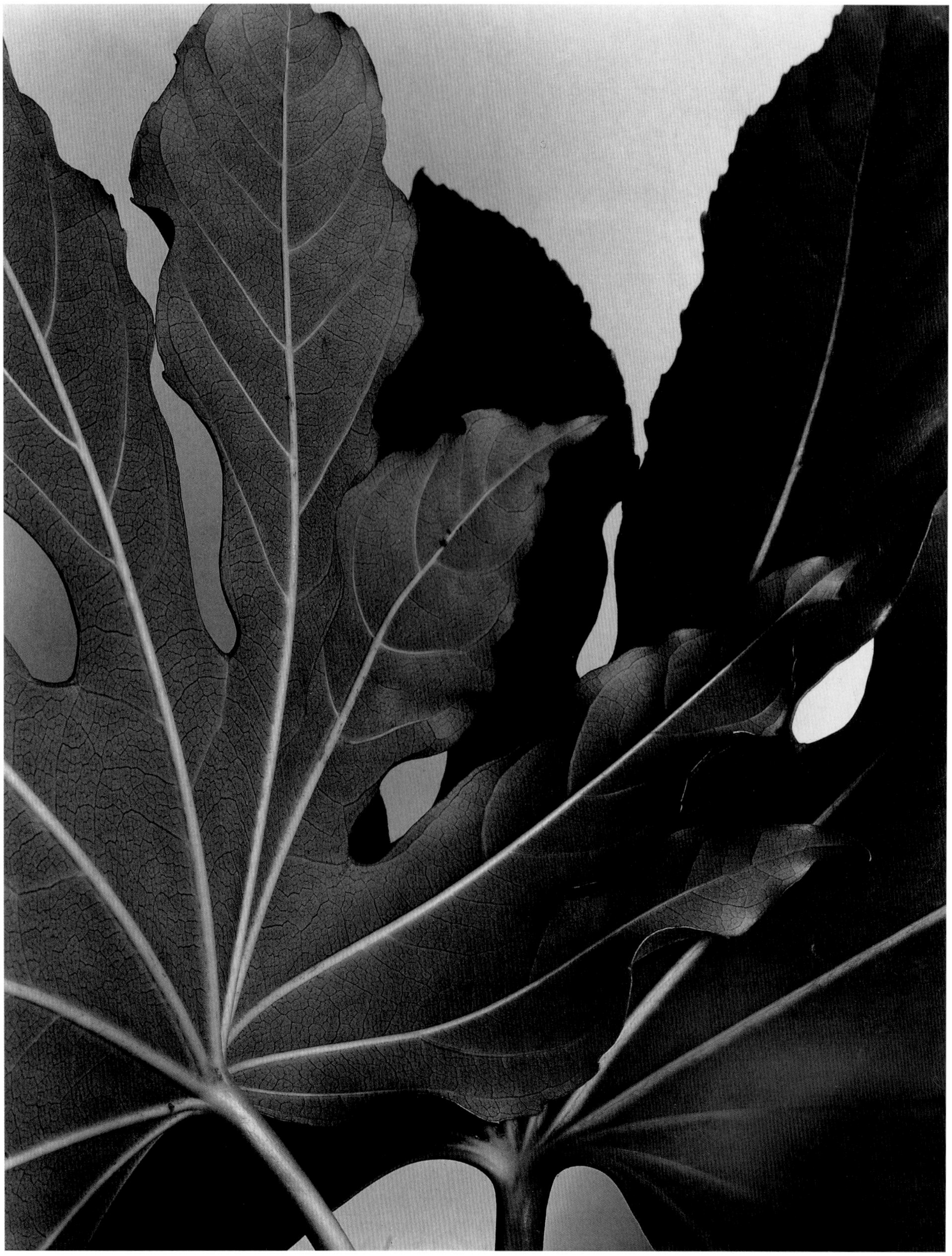

PLATE 64

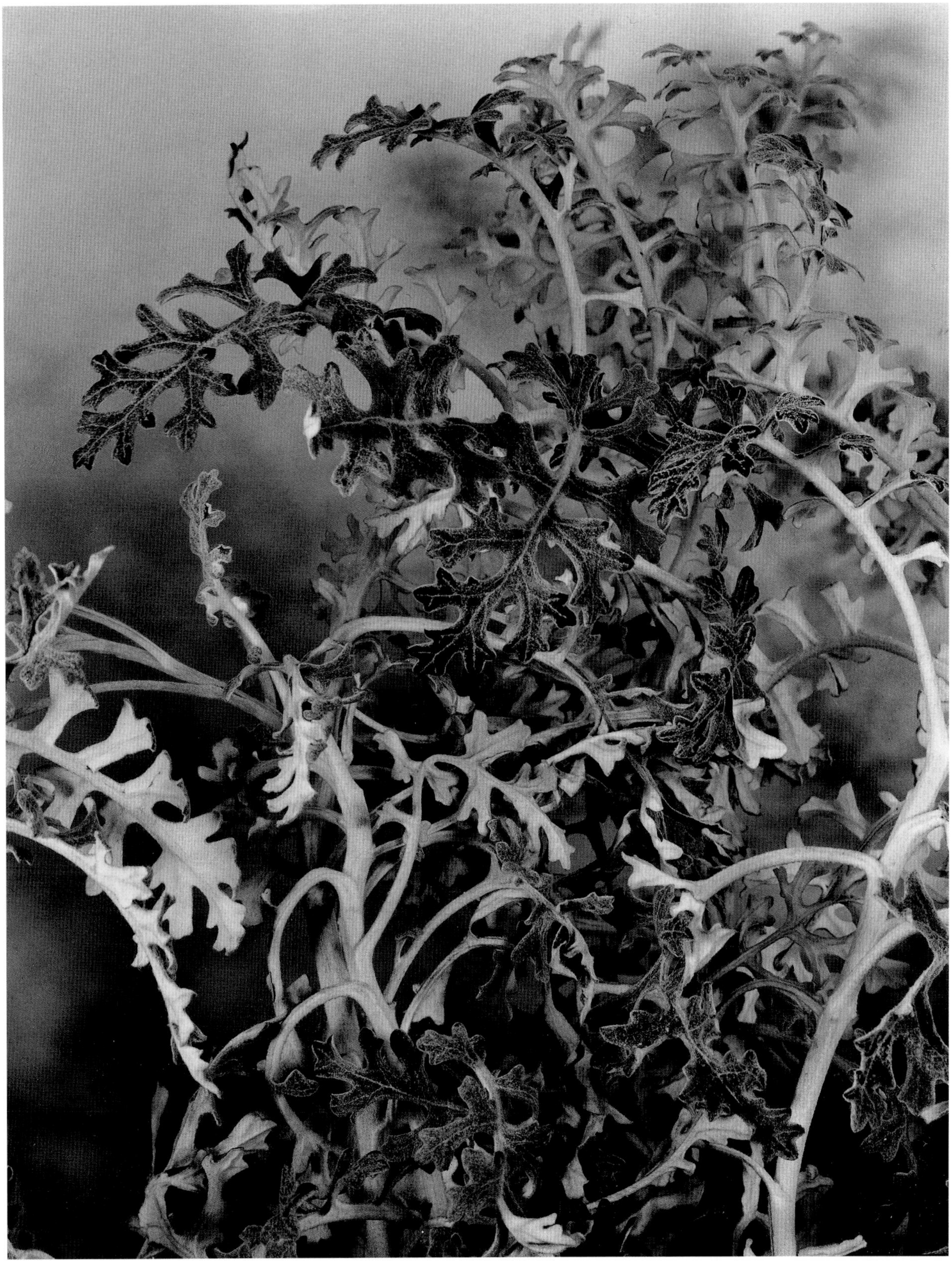

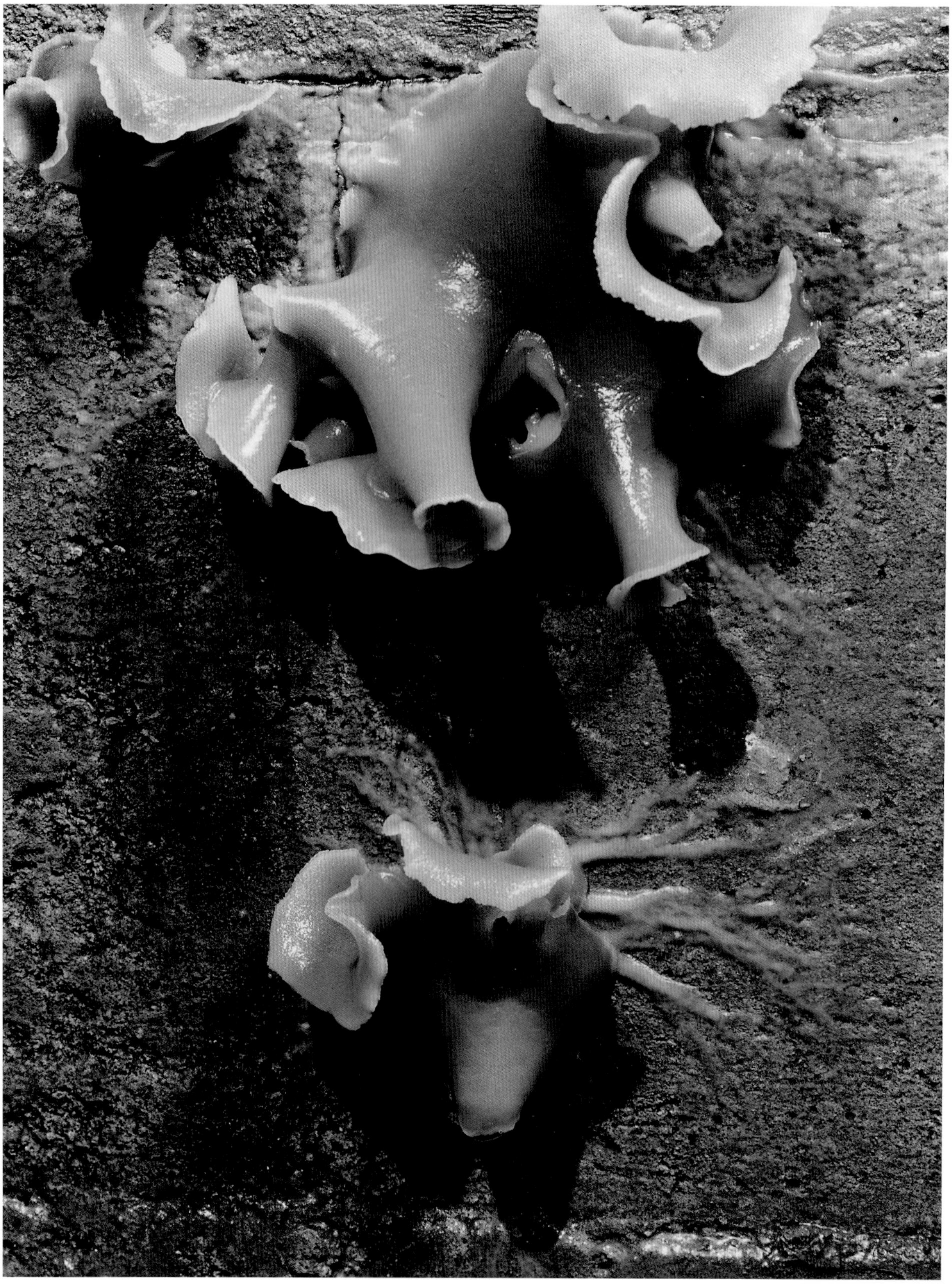

PLATE 66

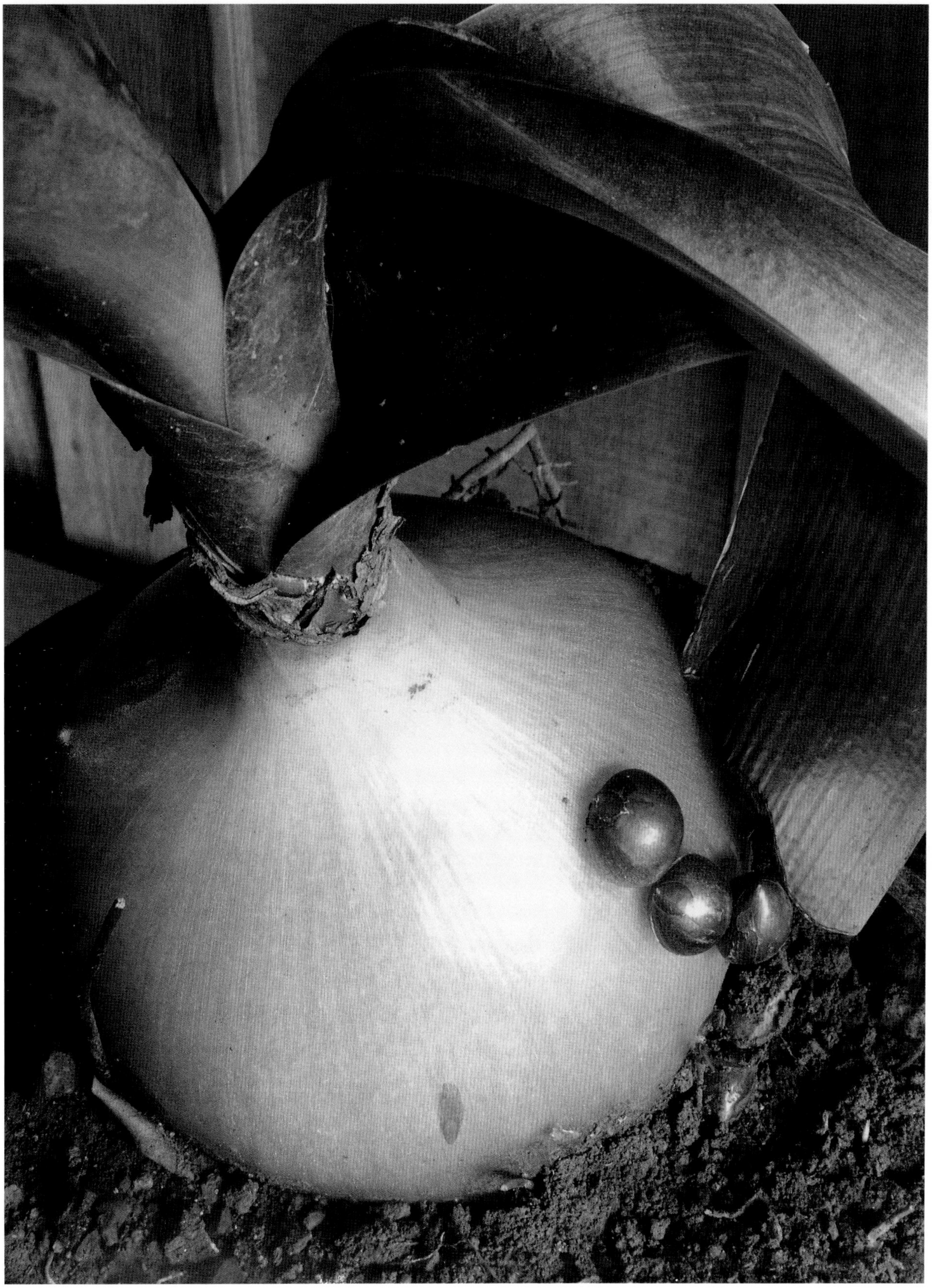

PLATE 67

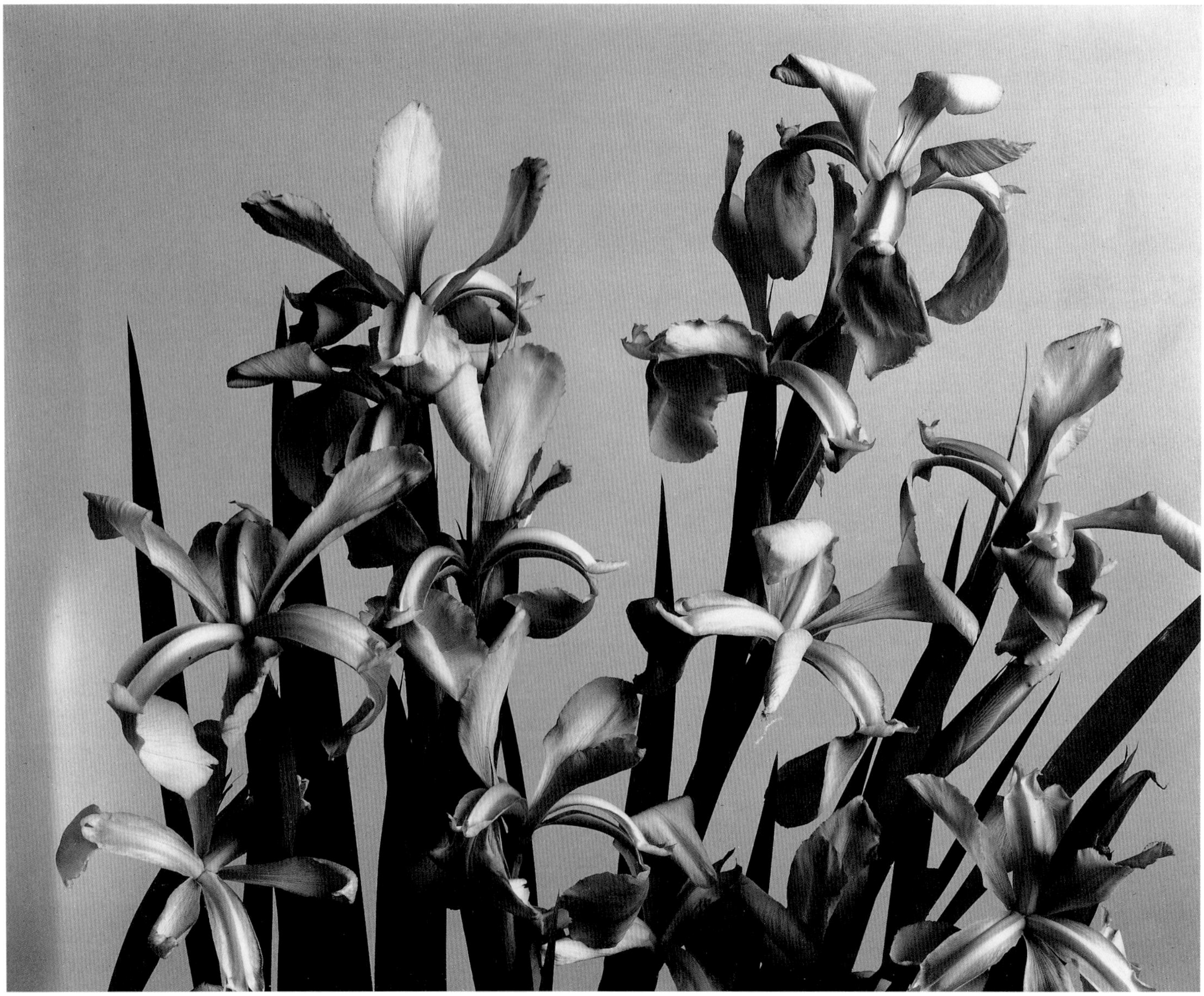

PLATE 68

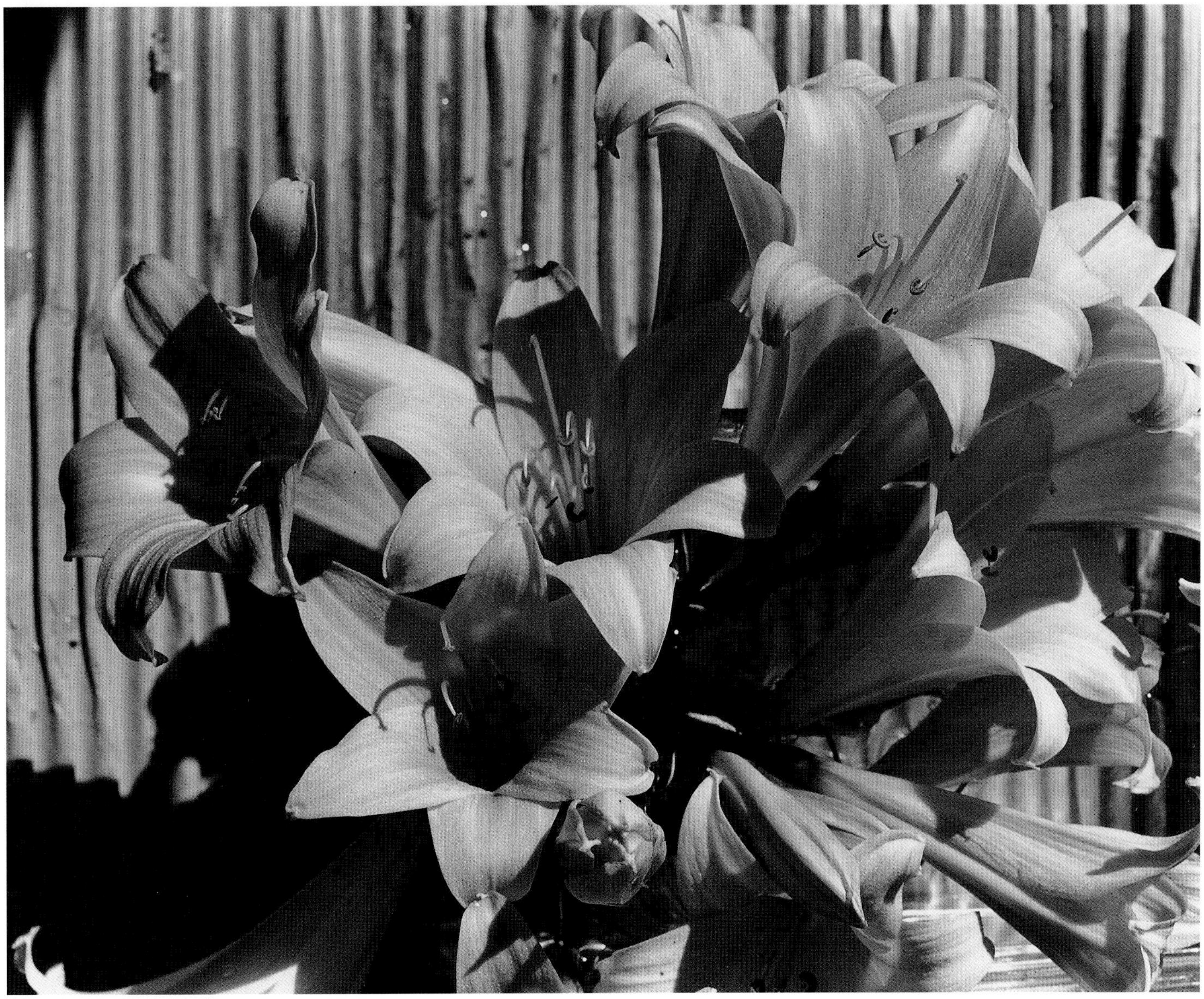

PLATE 69

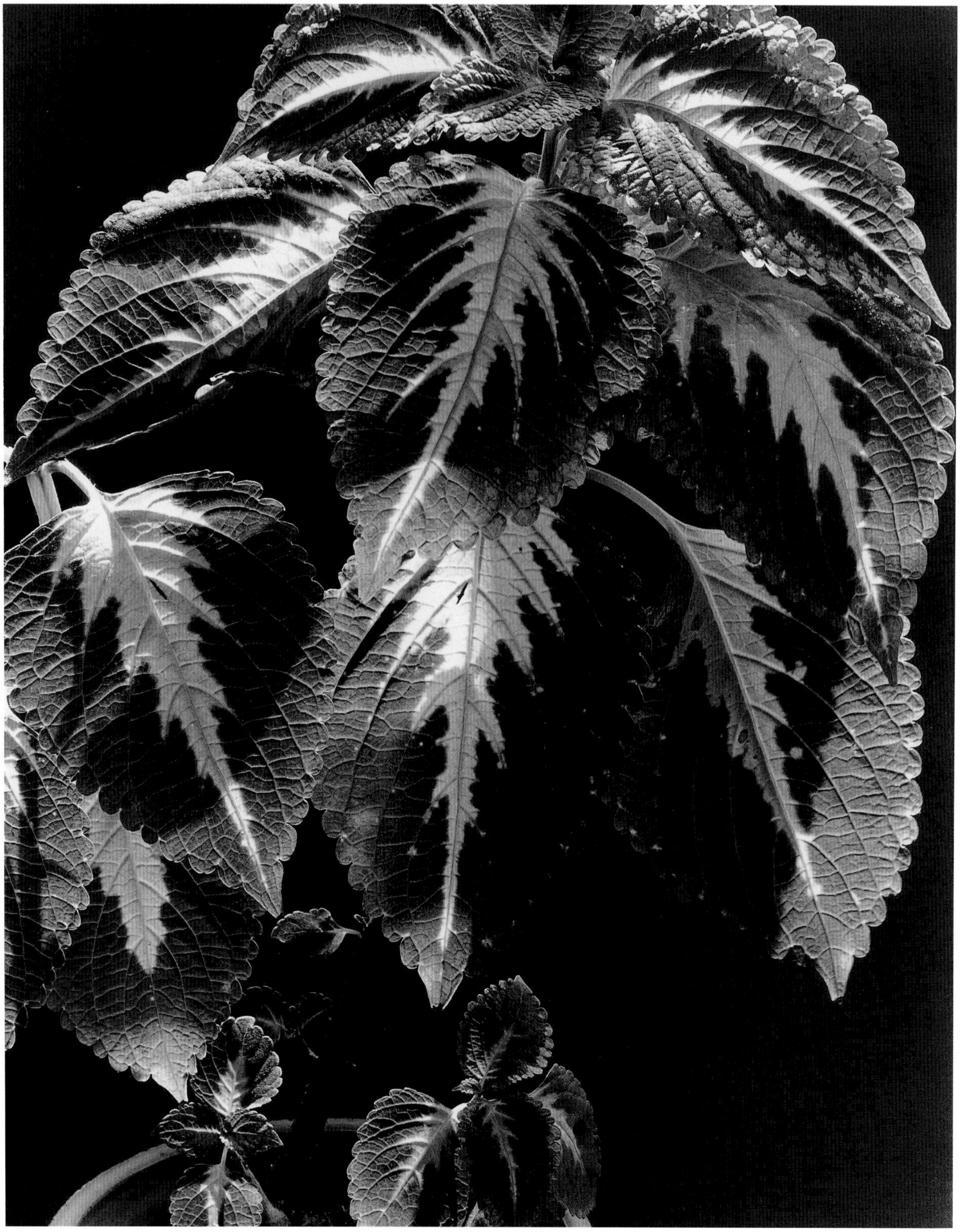

PLATE 70

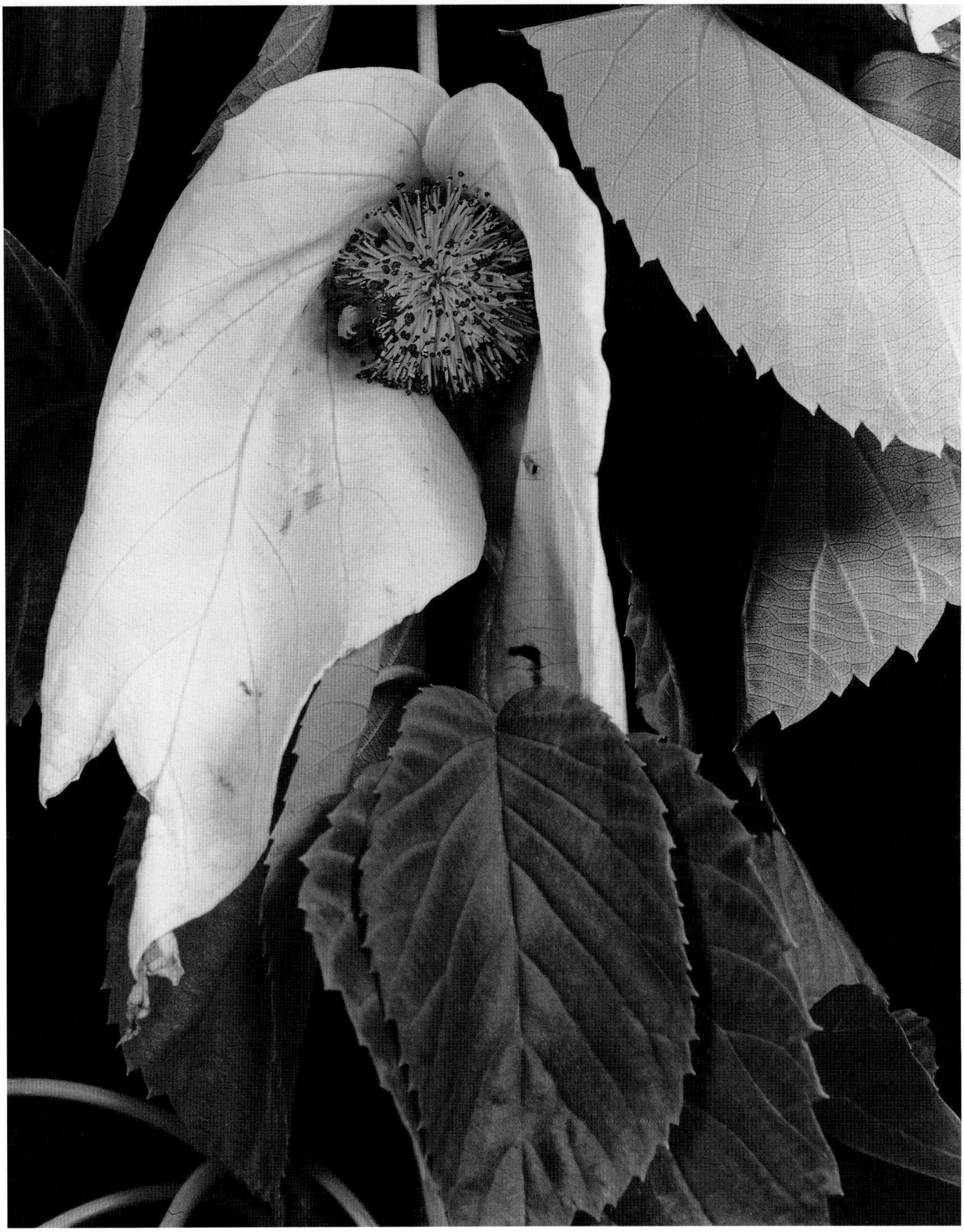

PLATE 71

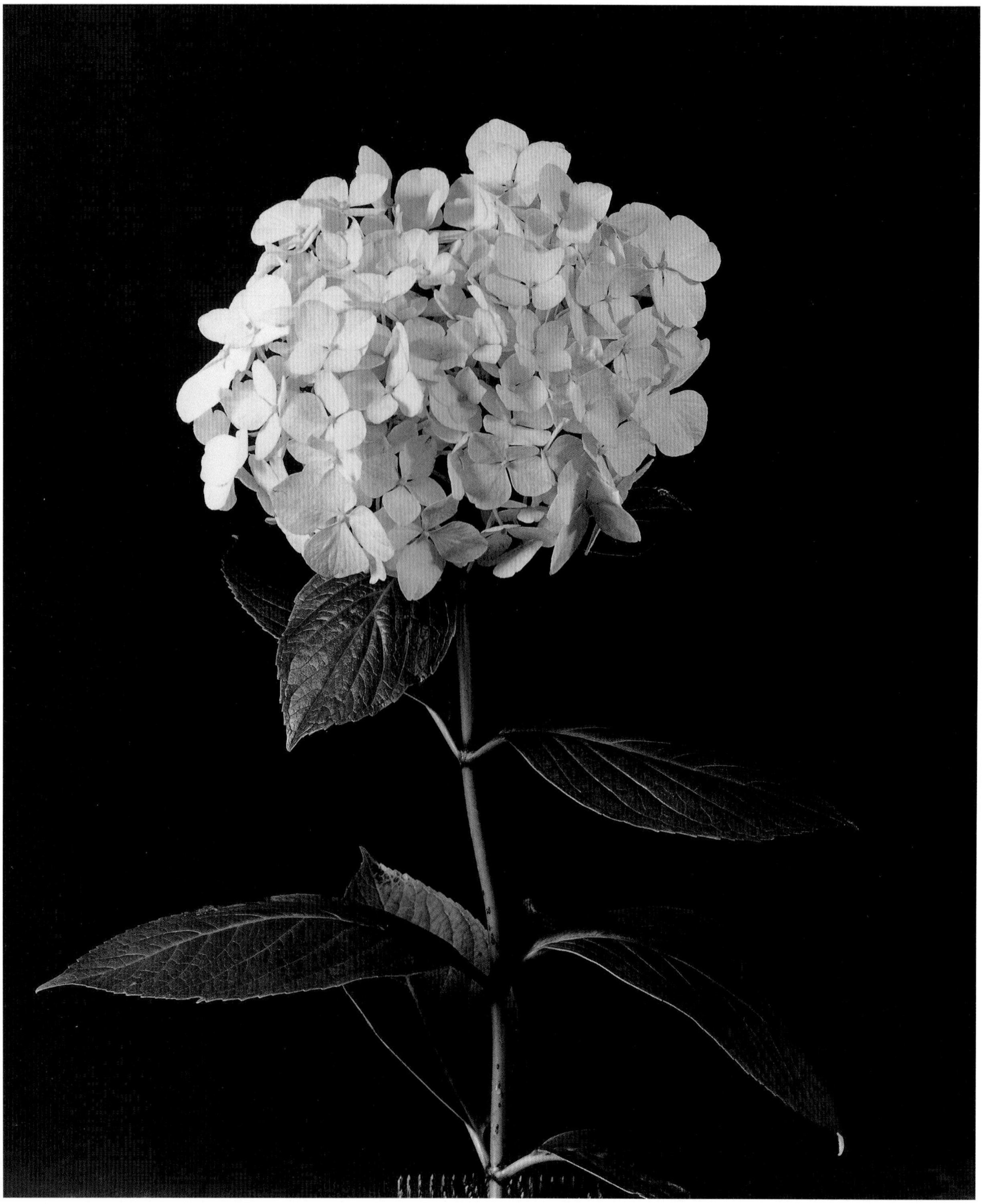

PLATE 72

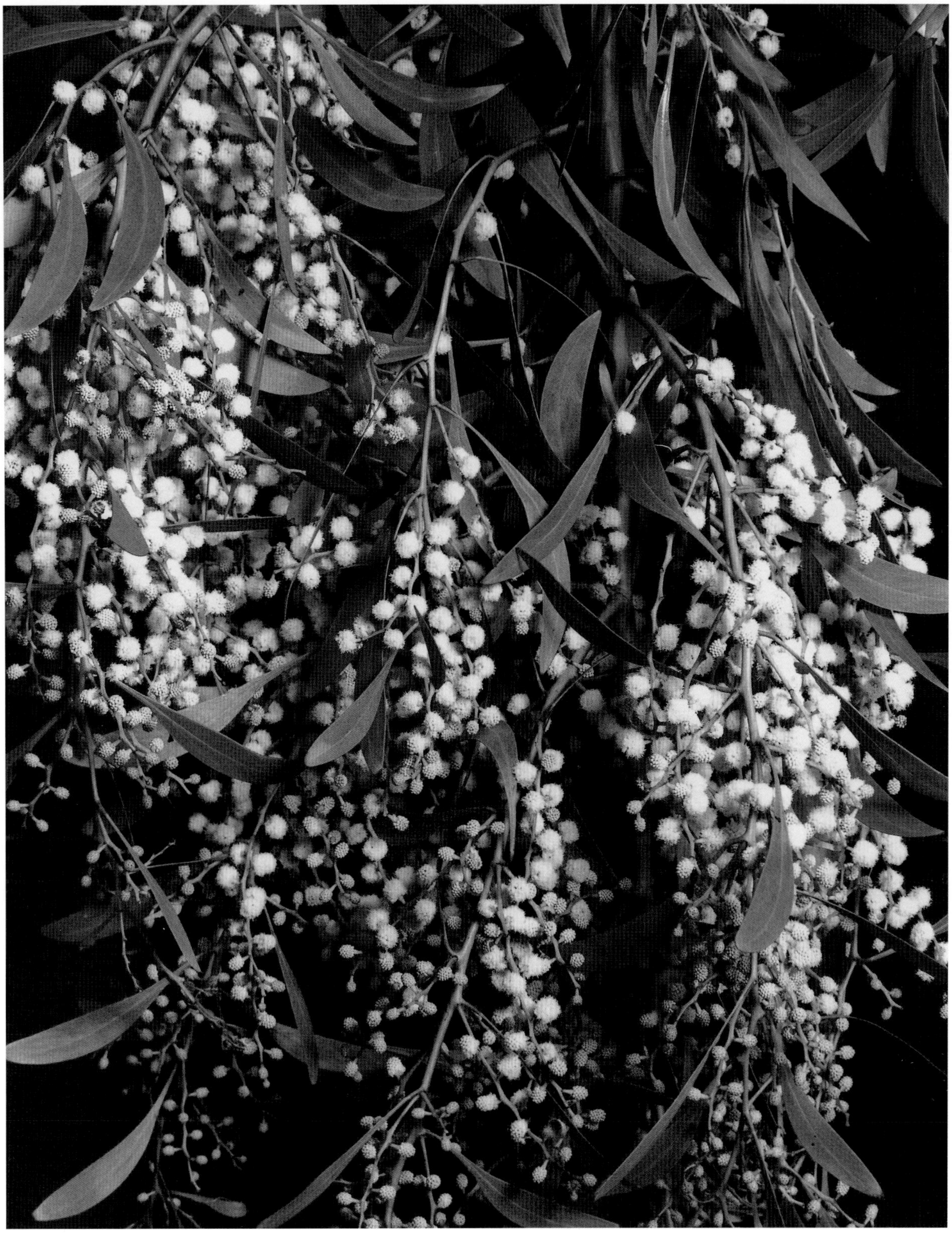

PLATE 73

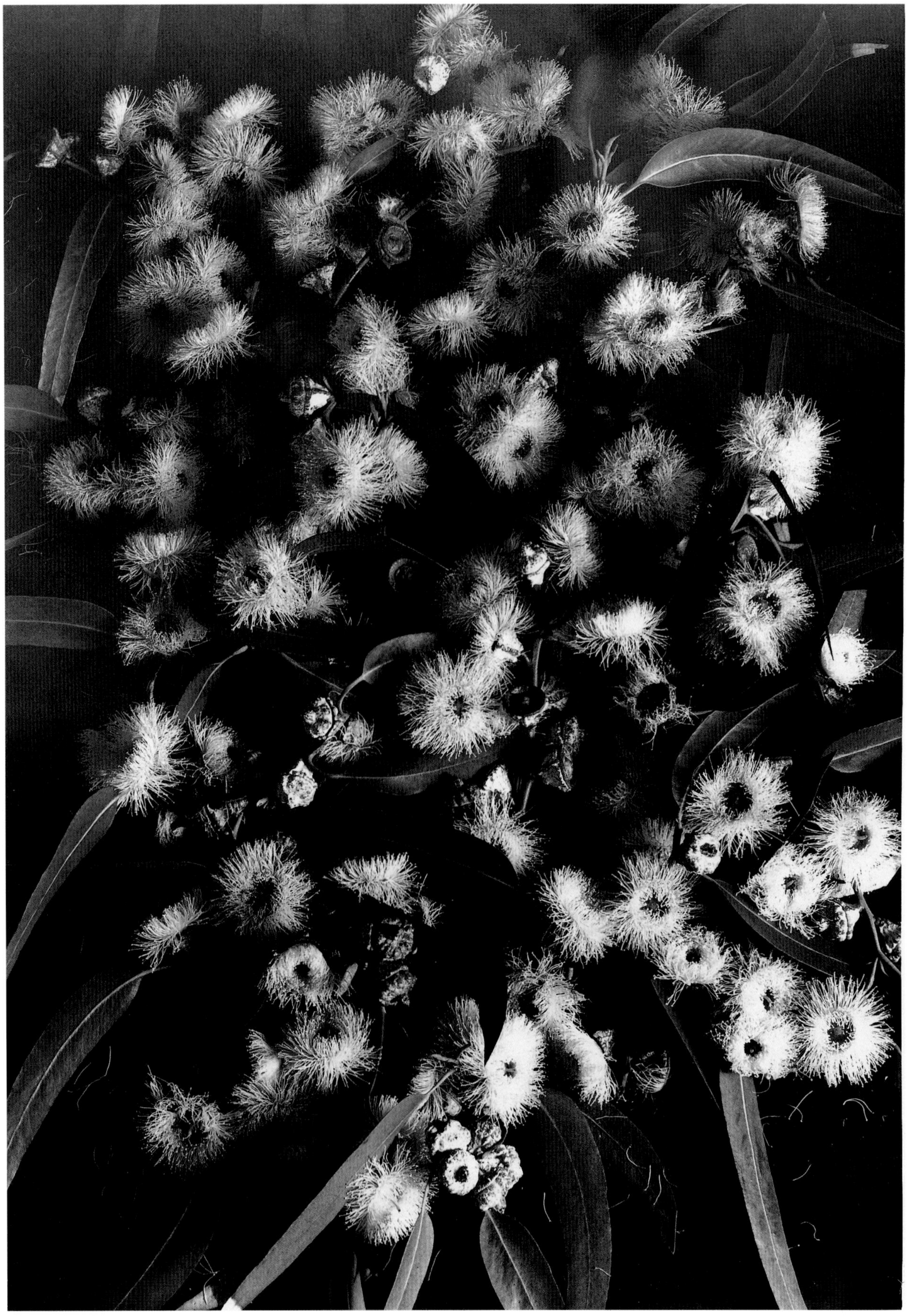

PLATE 74

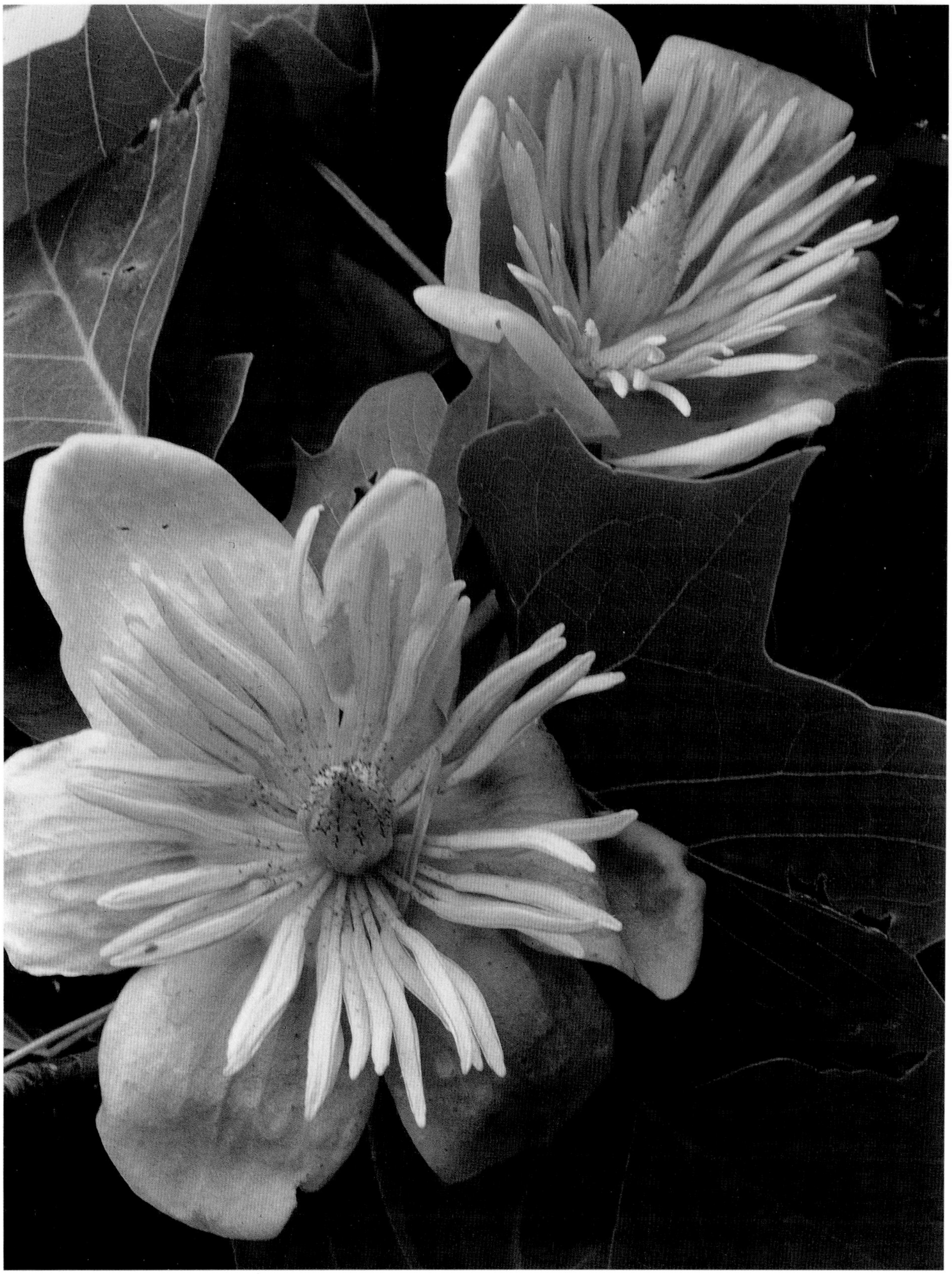

PLATE 75

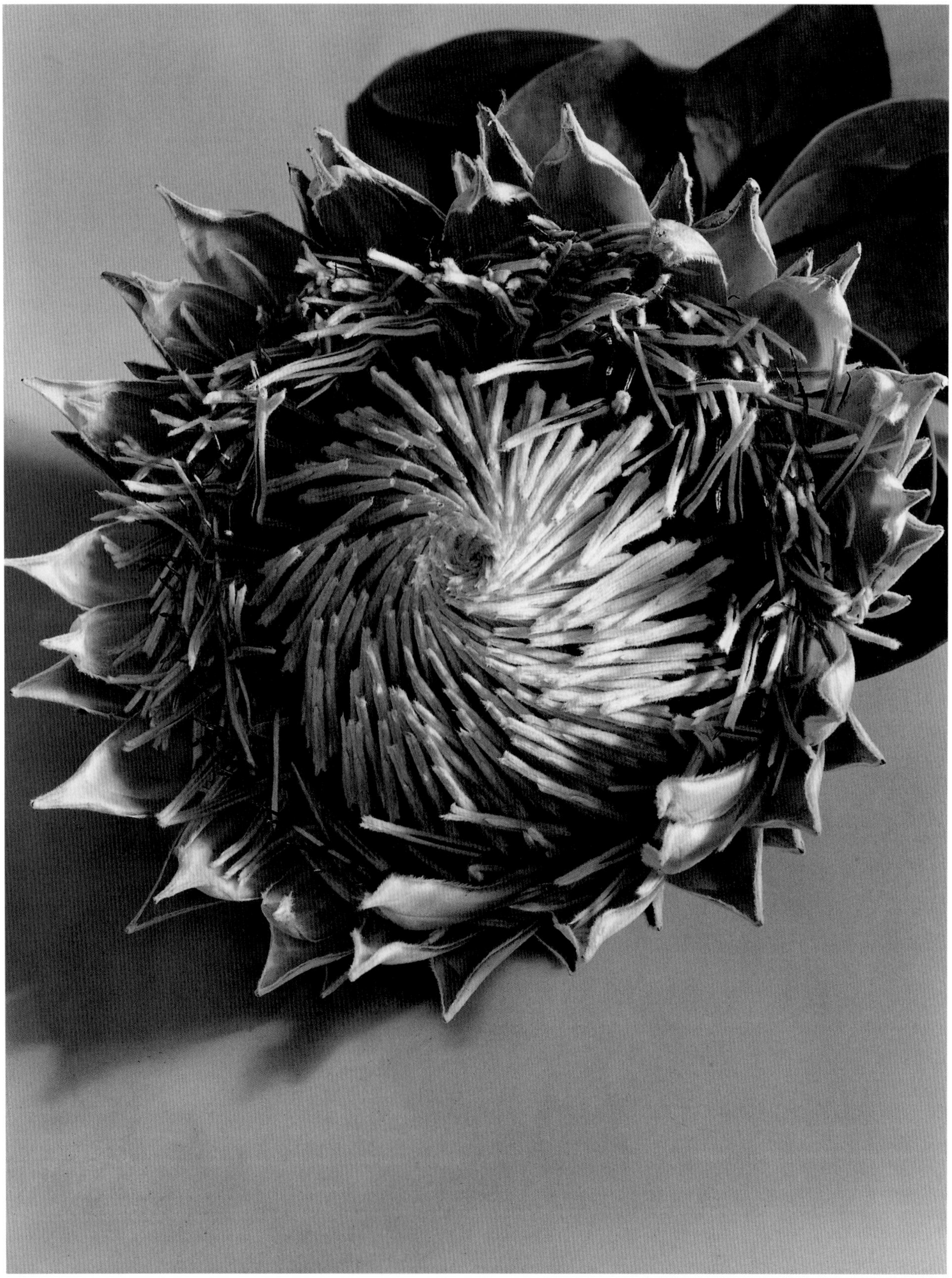

PLATE 76

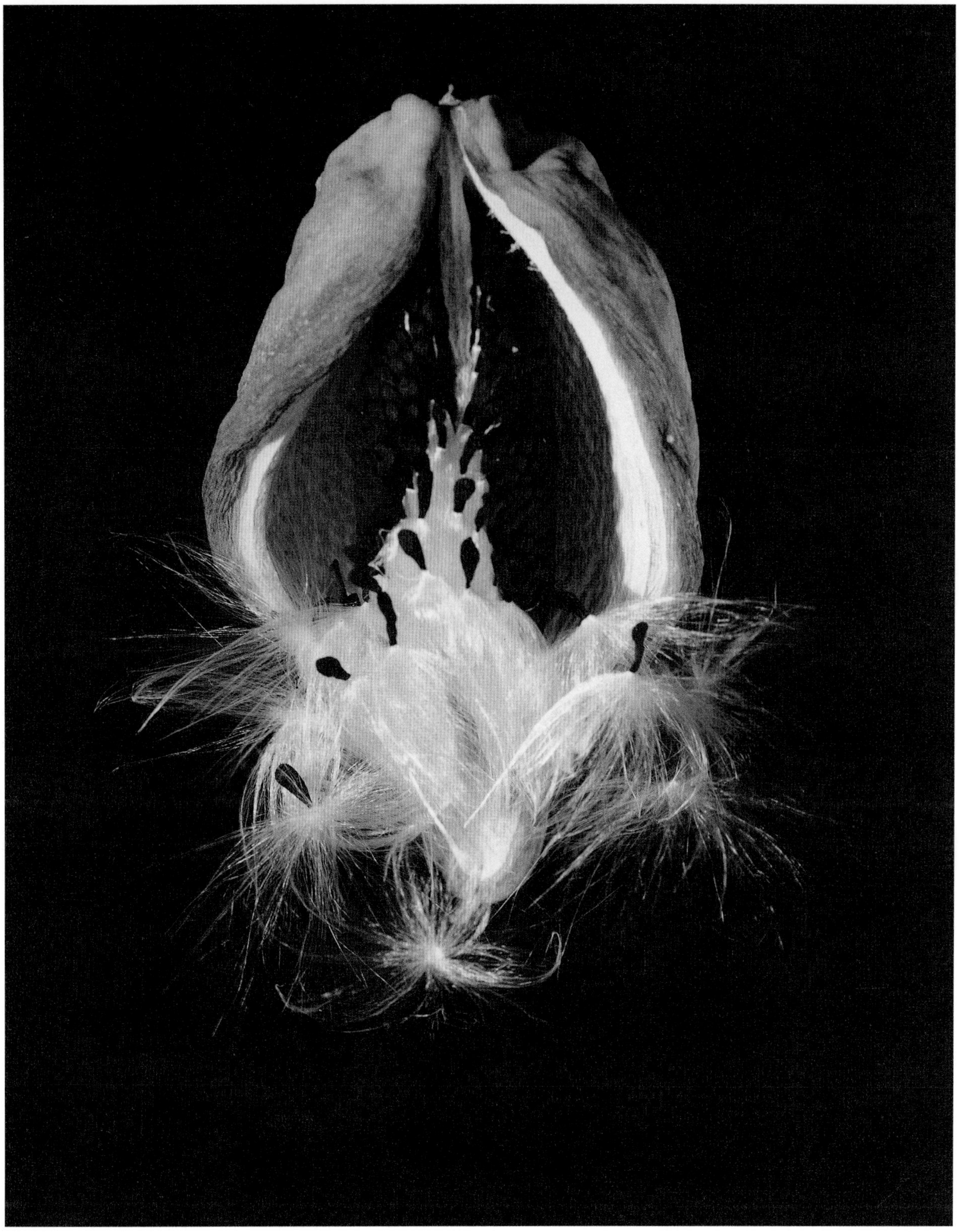

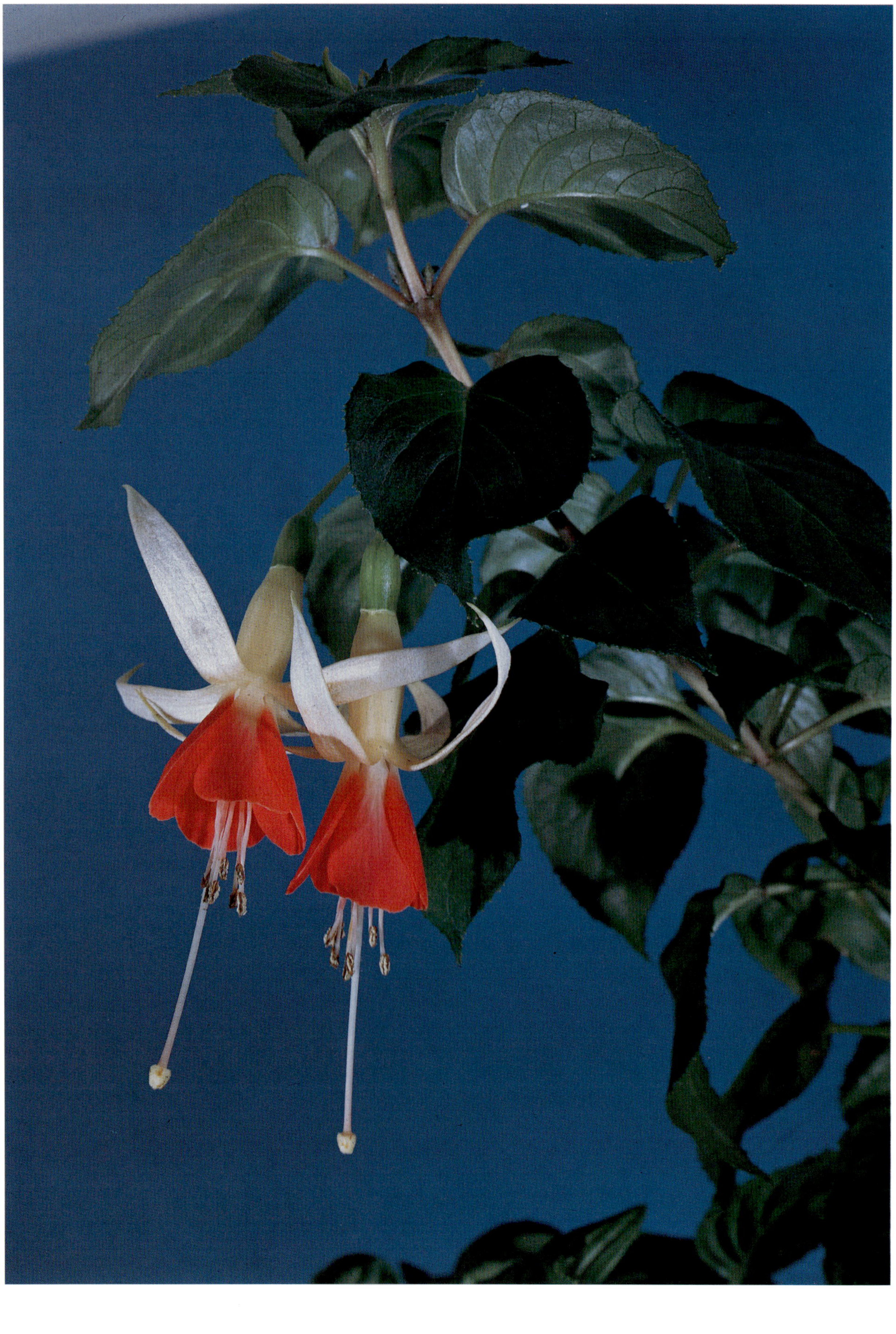

PLATE 78

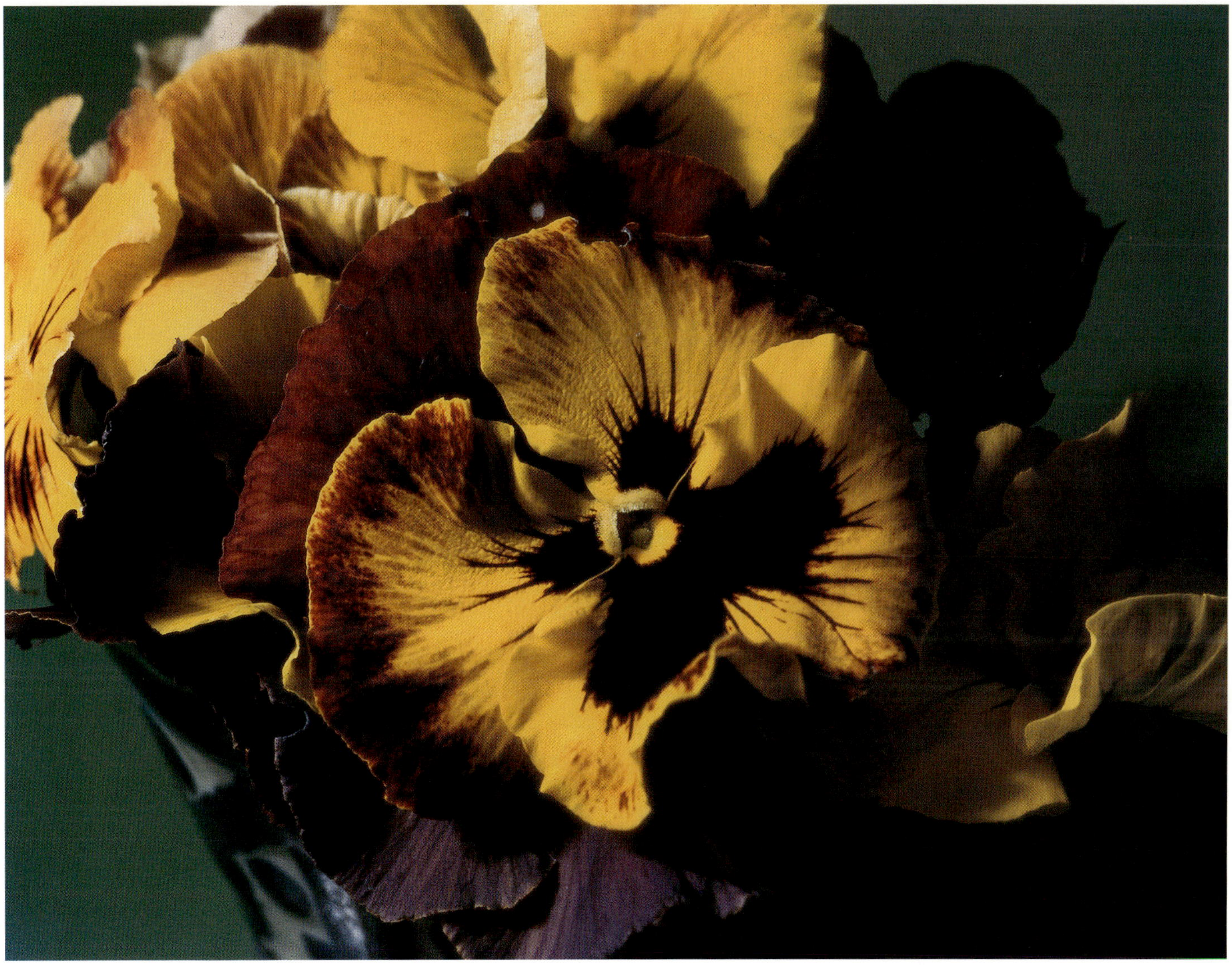

PLATE 79

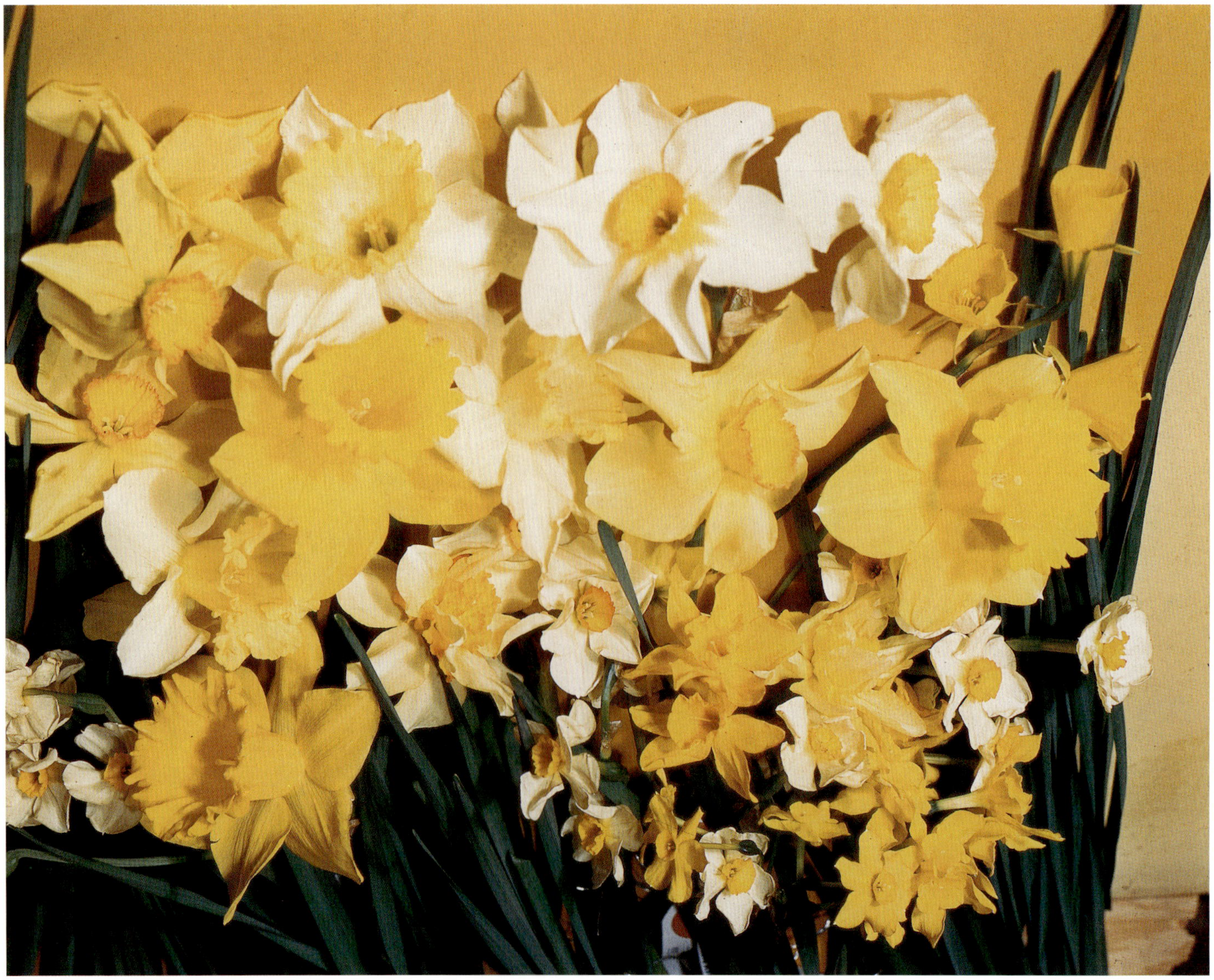

PLATE 80

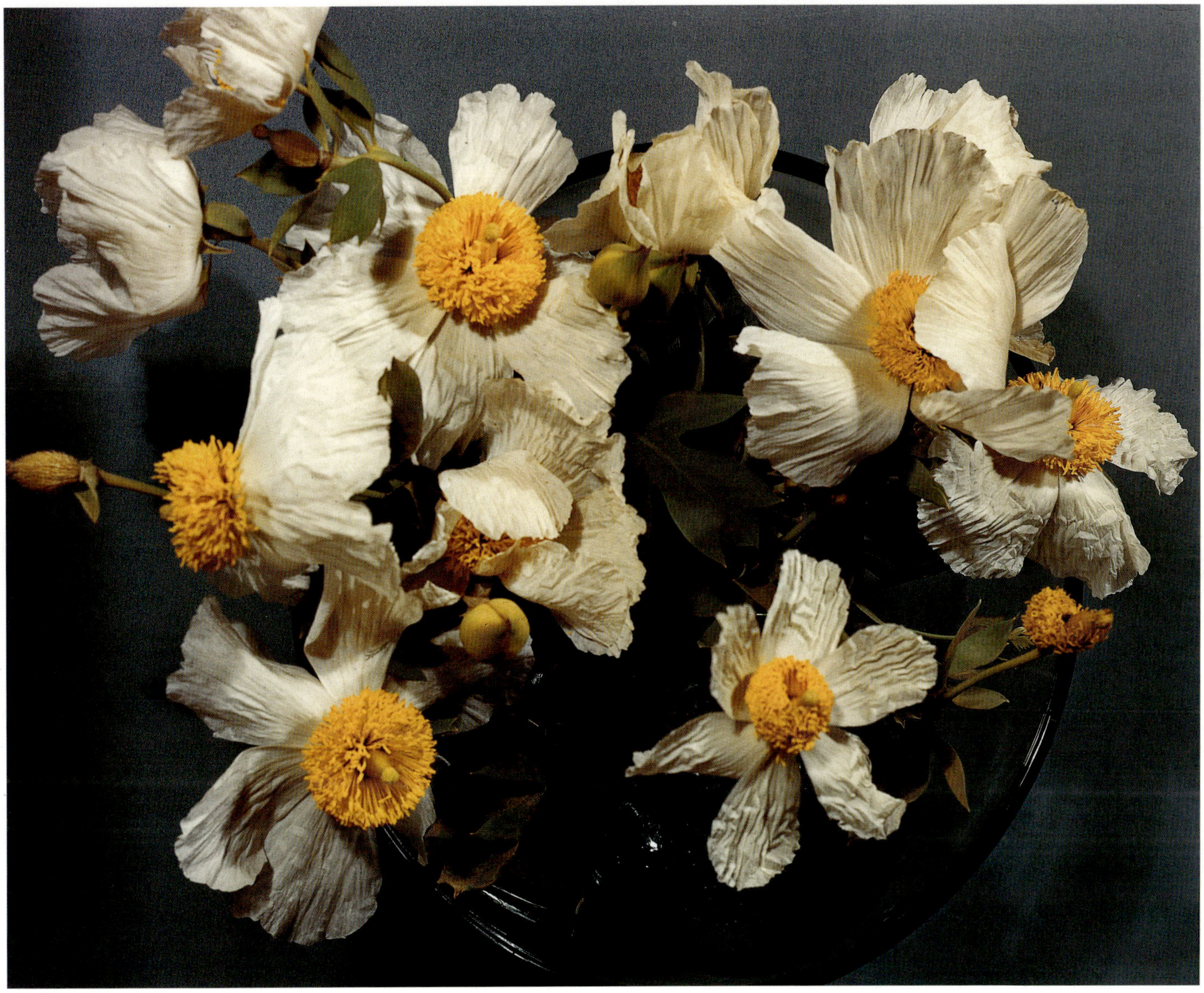

PLATE 81

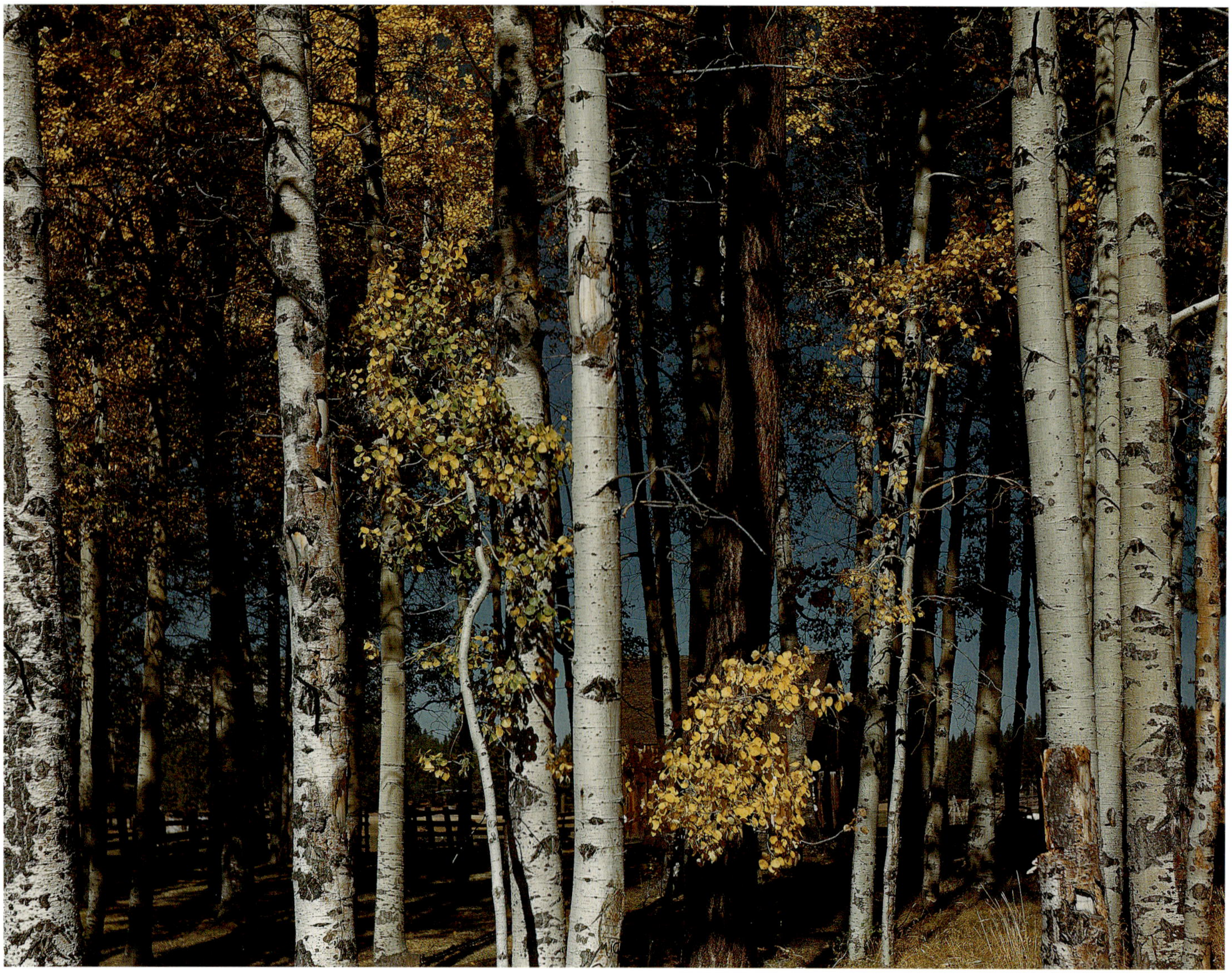

PLATE 82

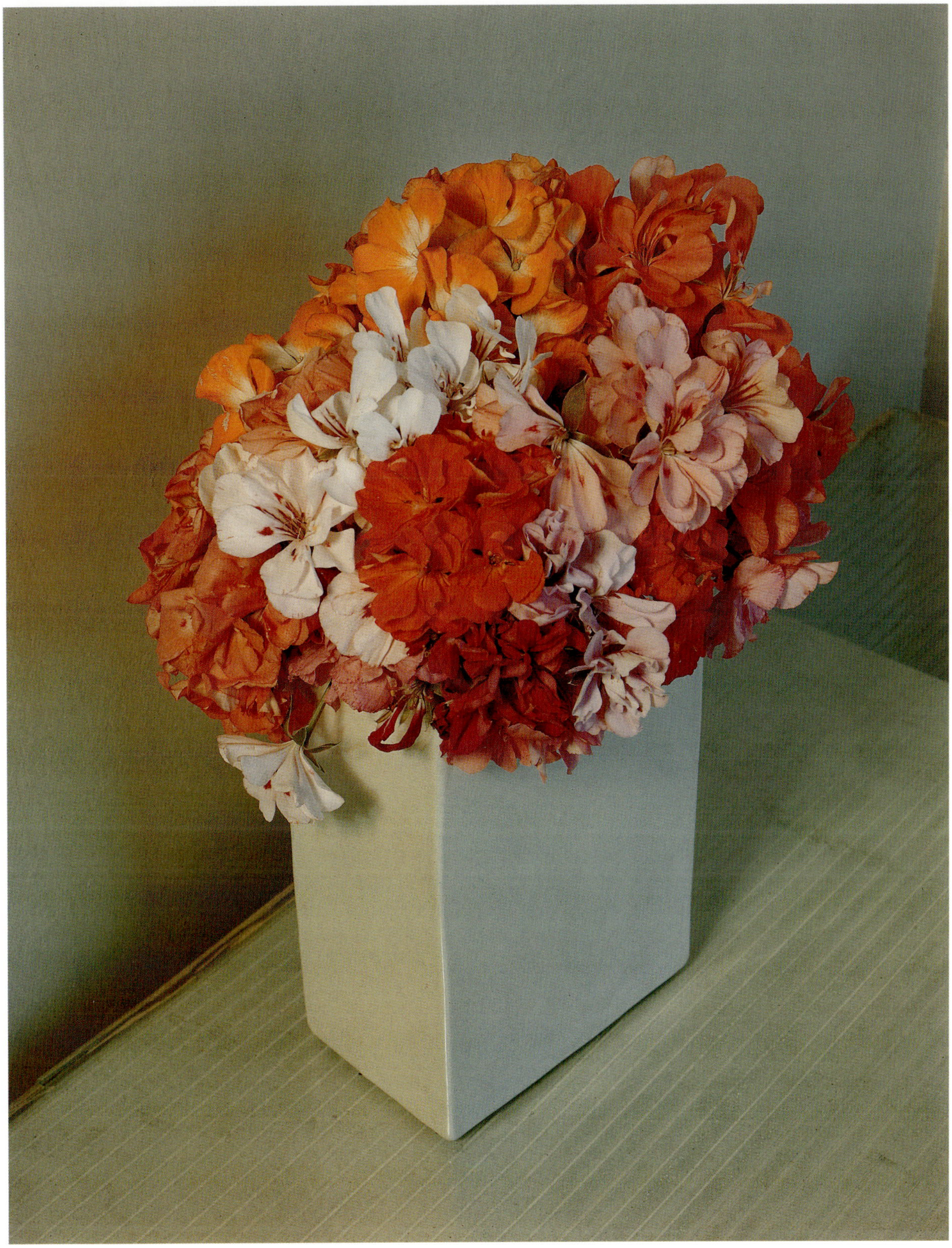

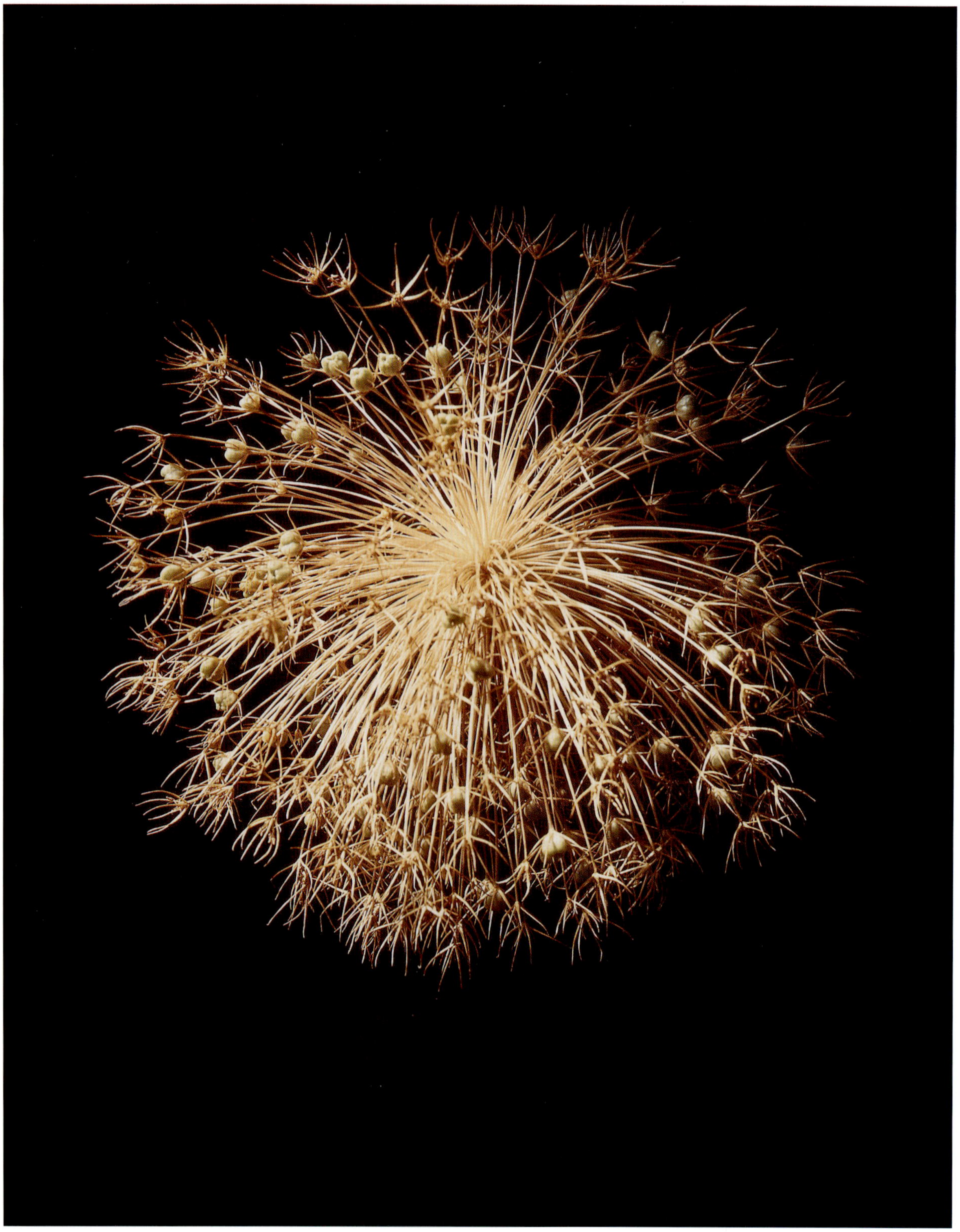

PLATE 84

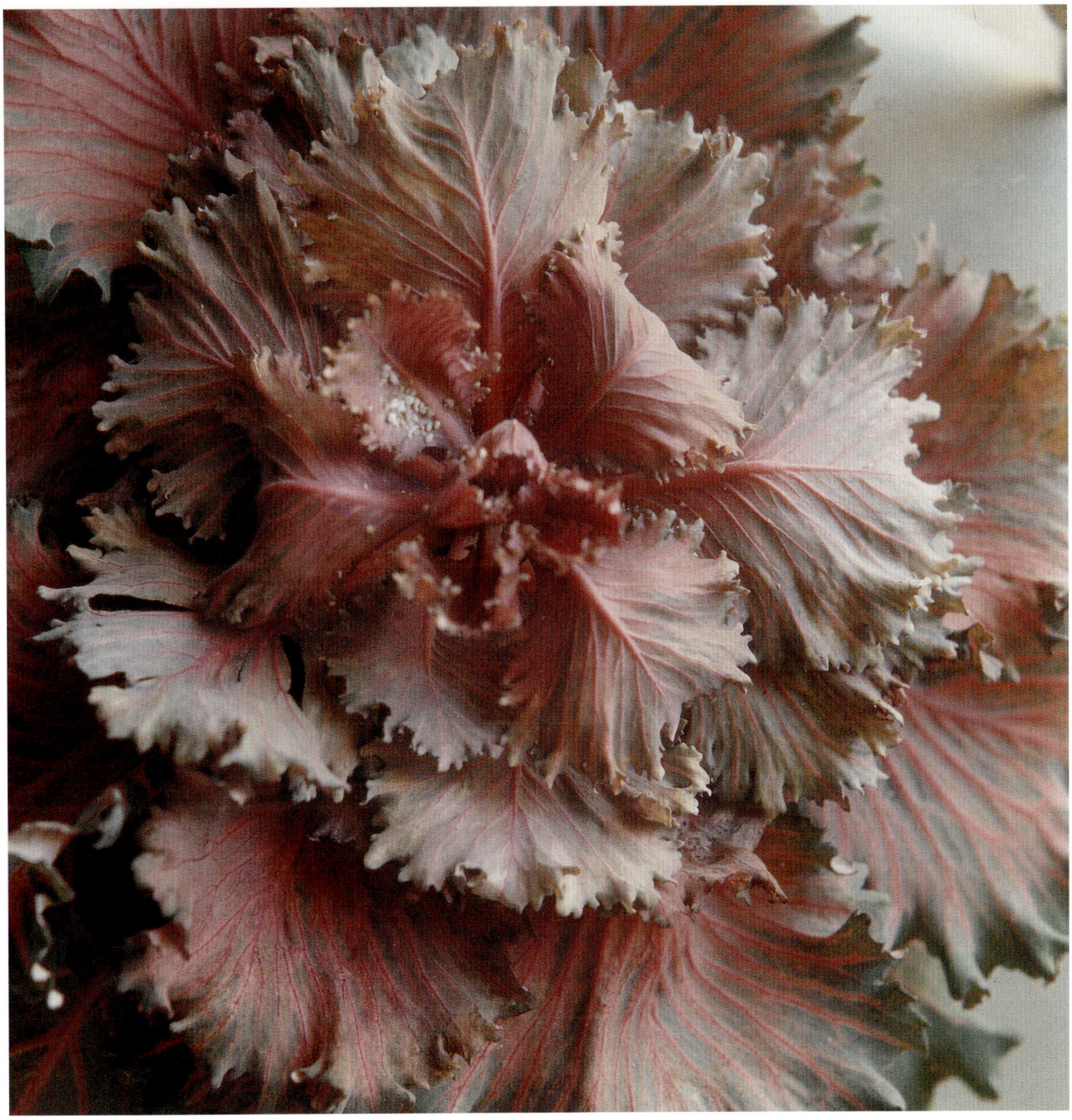

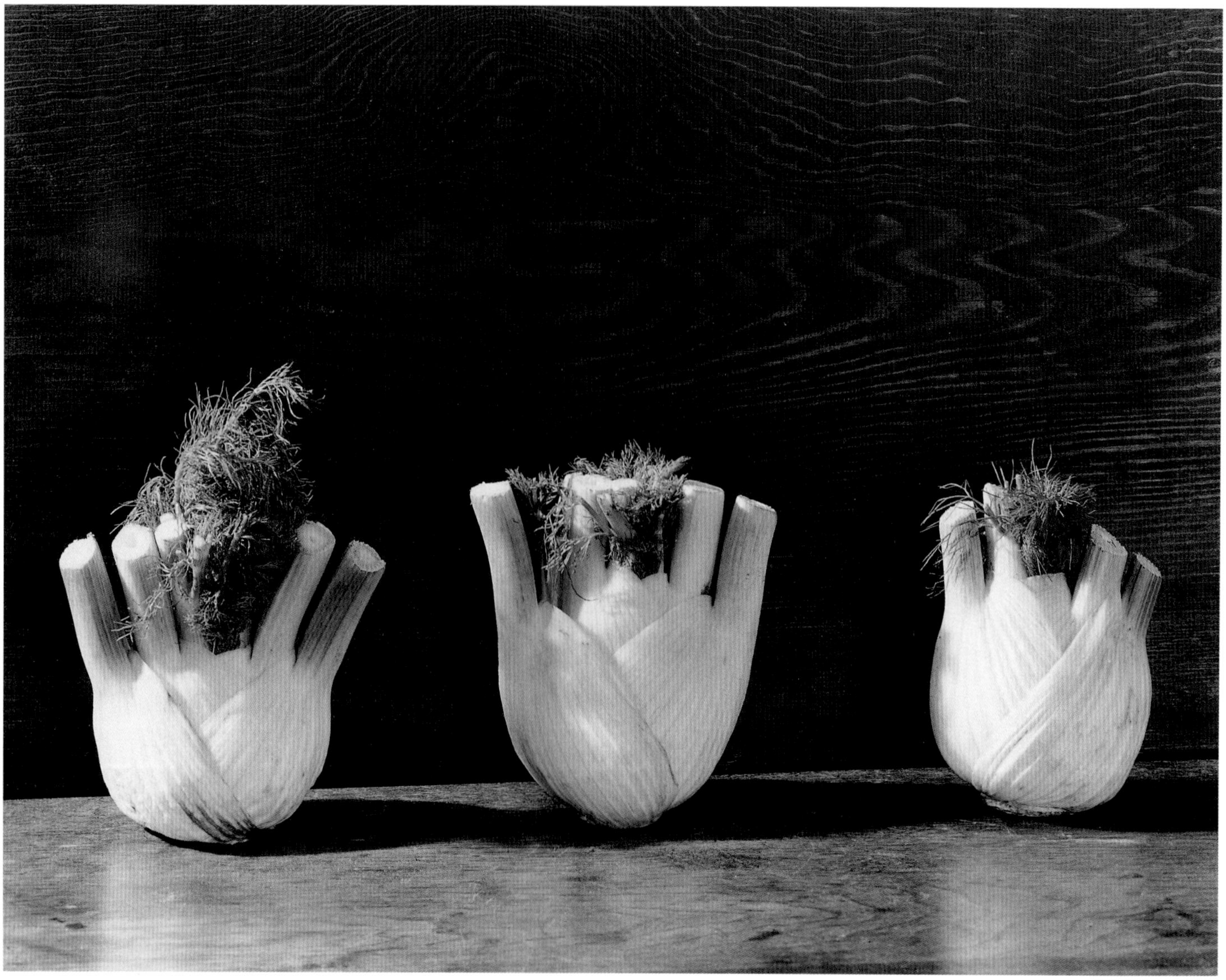

PLATE 86

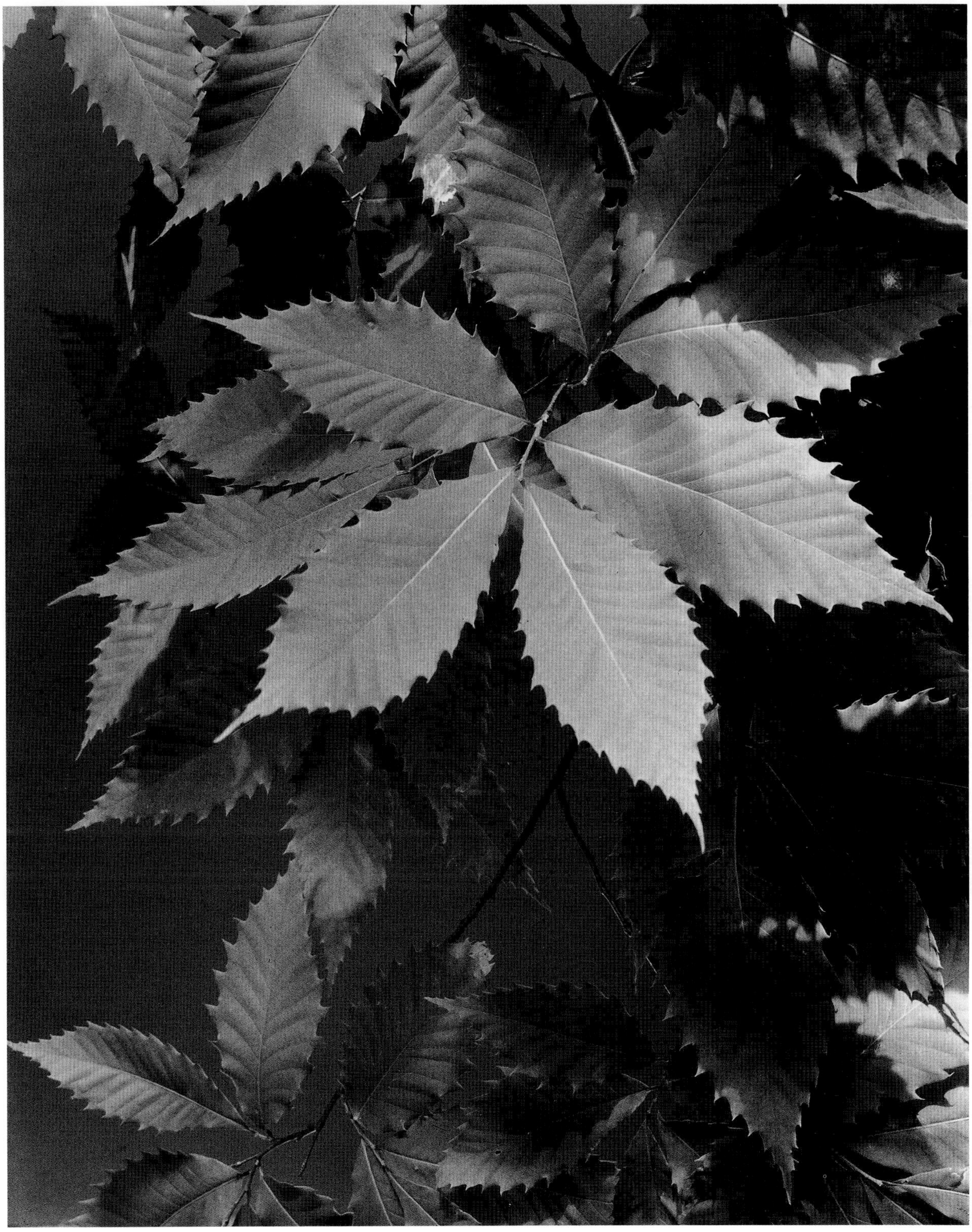

PLATE 87

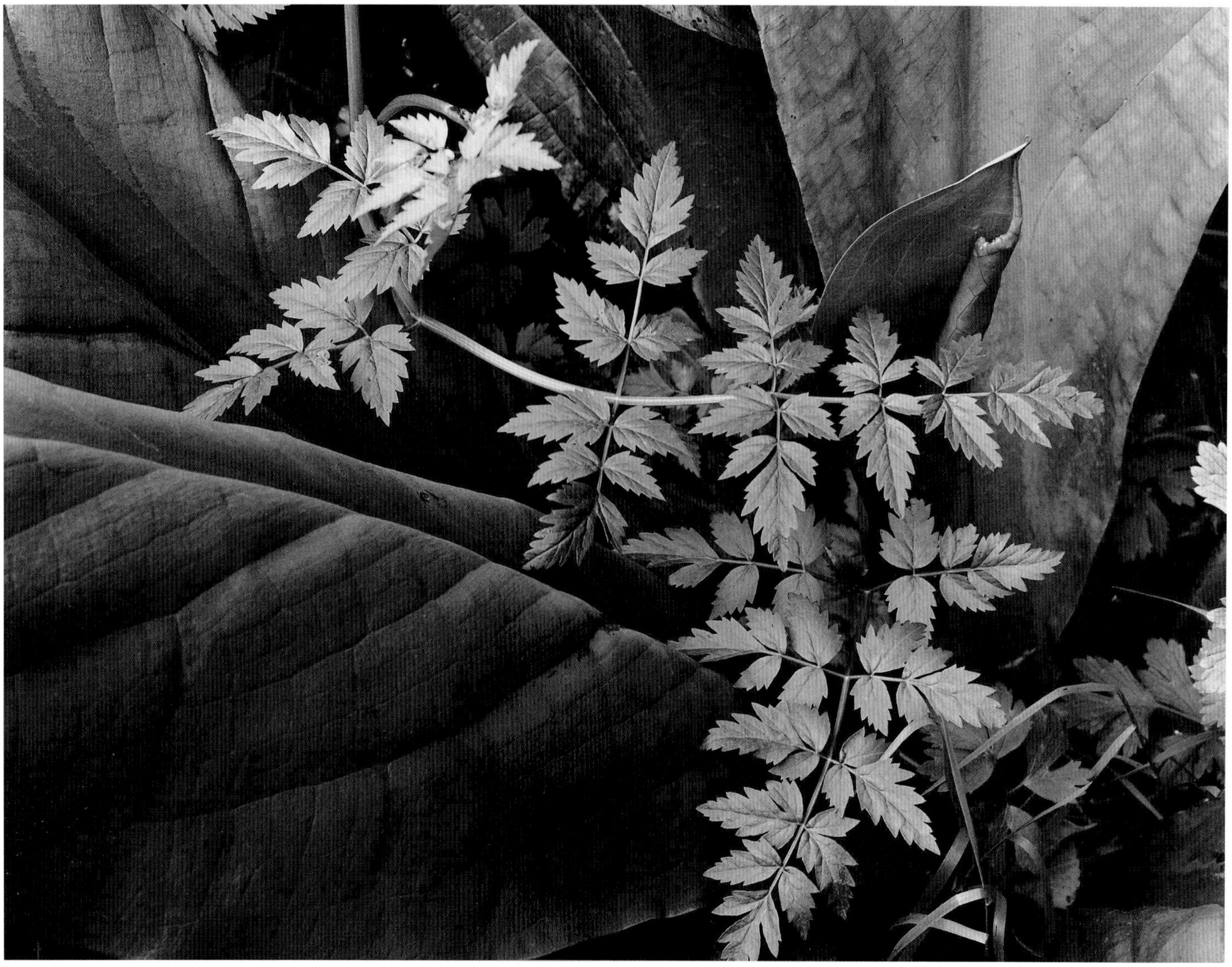

PLATE 88

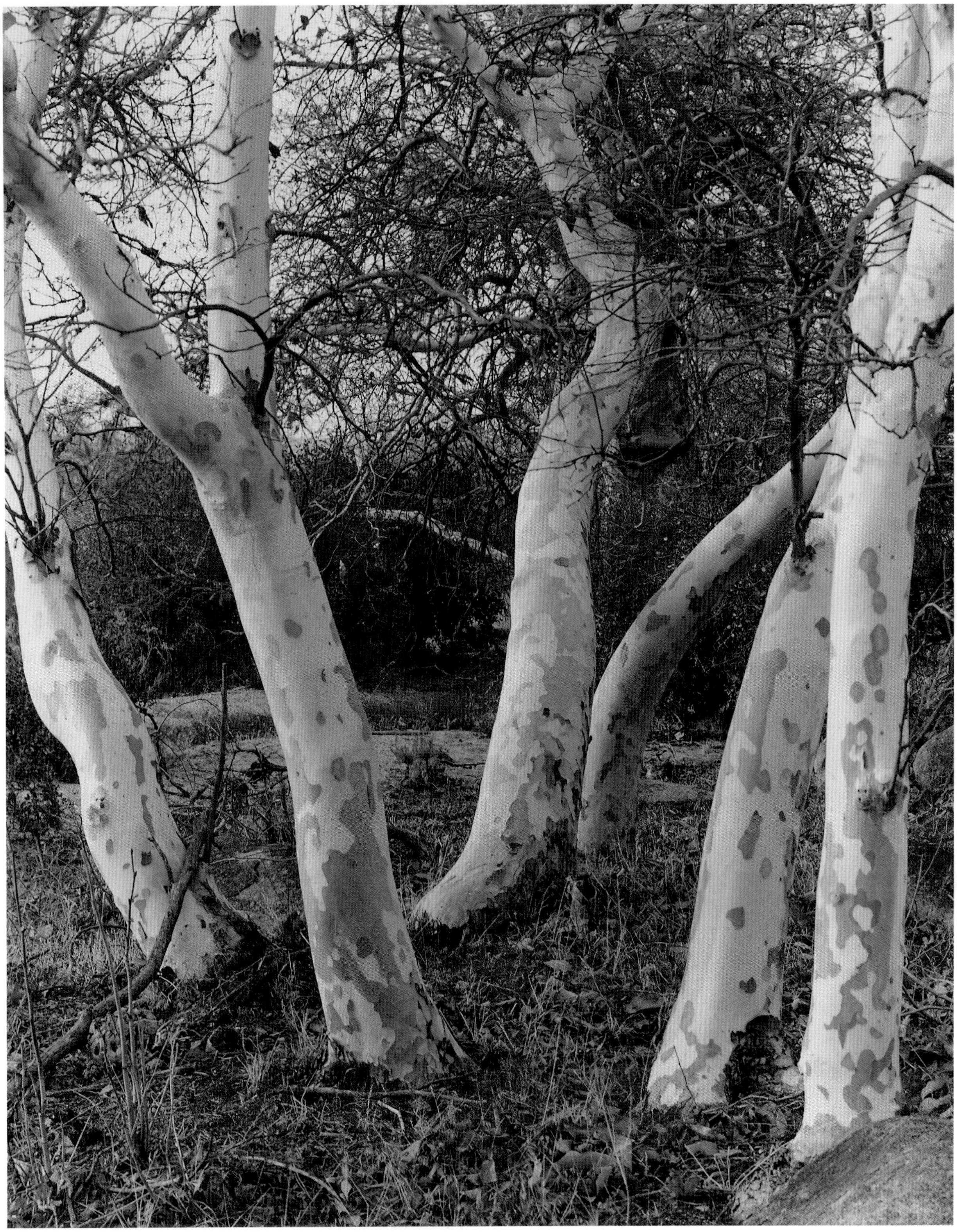

PLATE 89

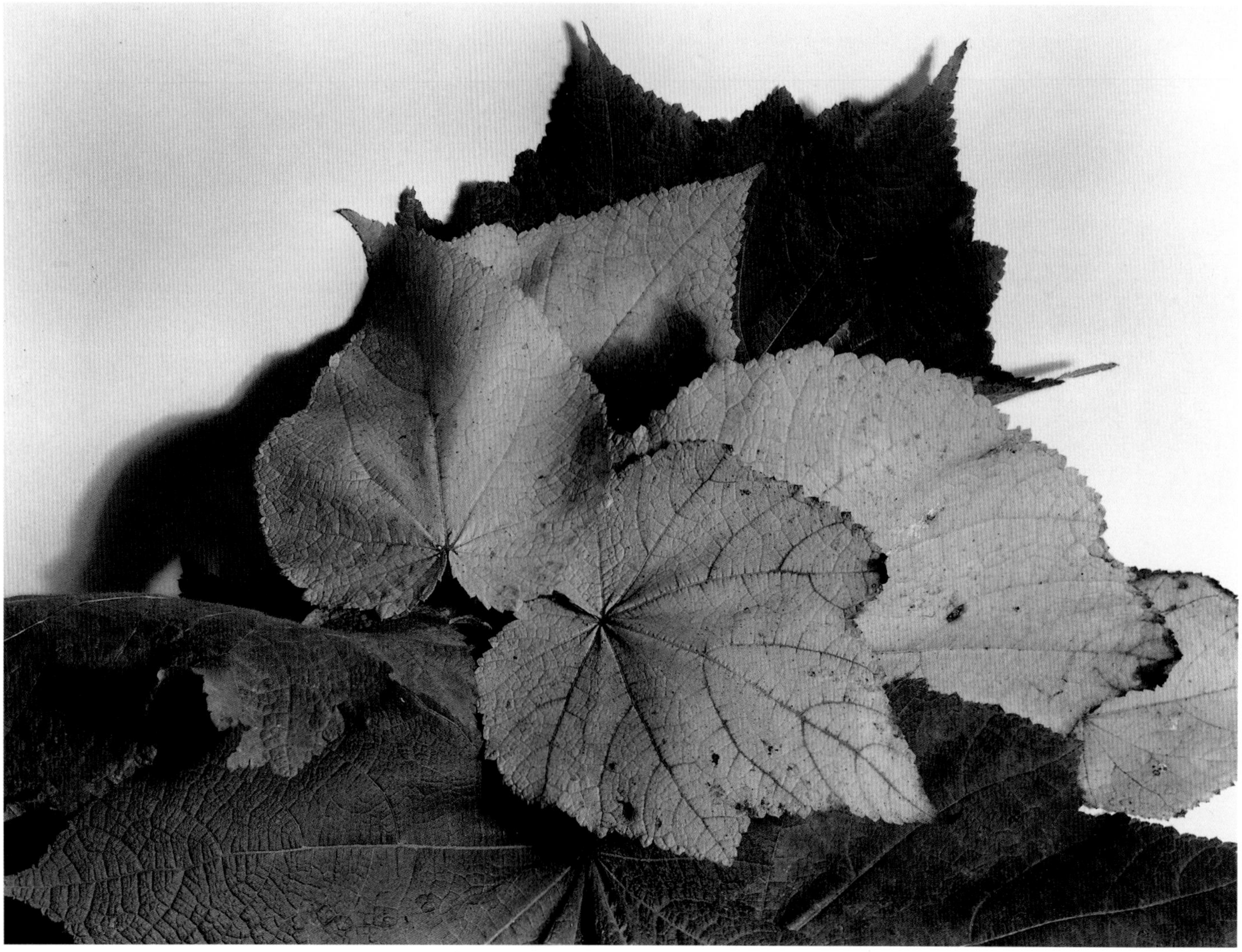

PLATE 90

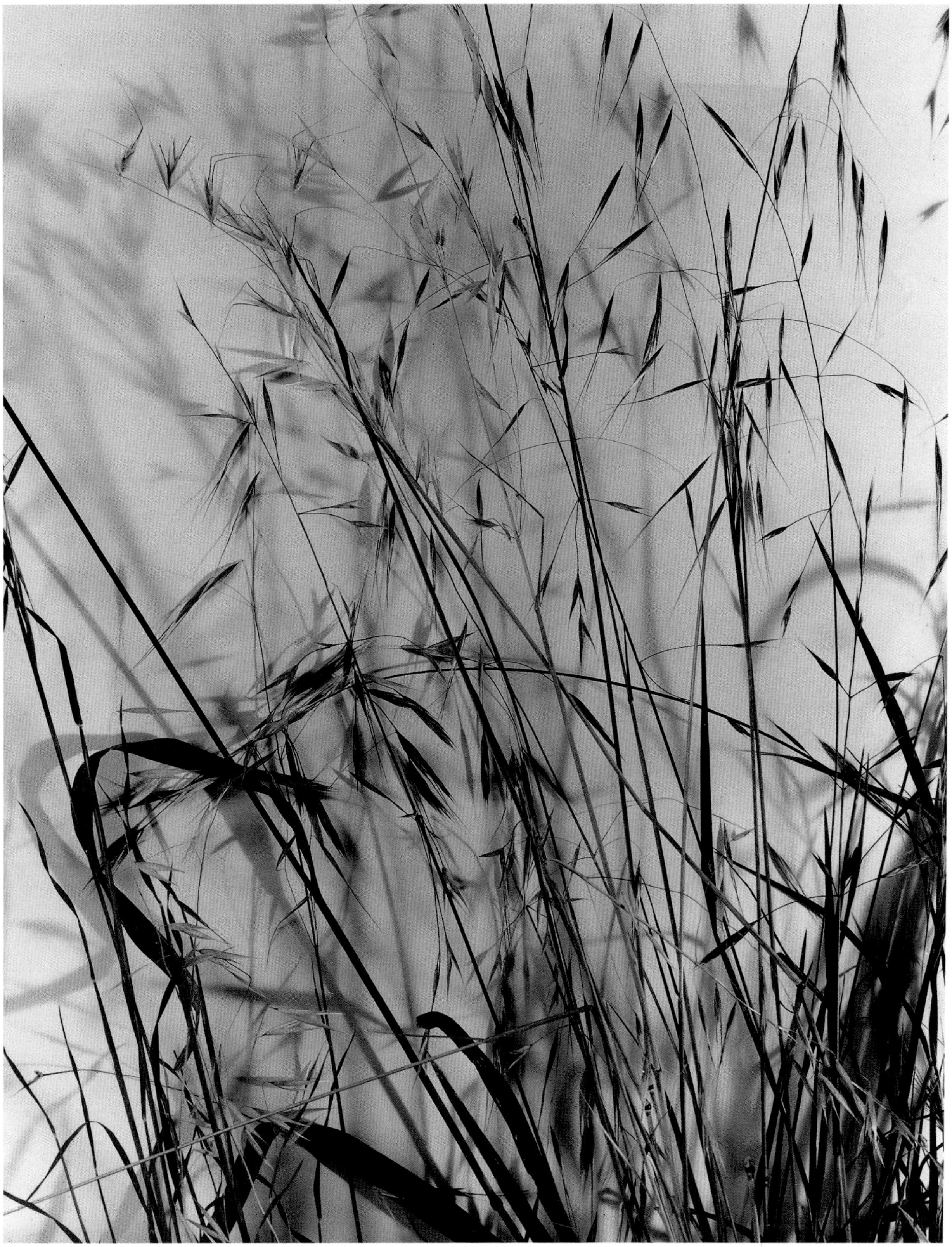

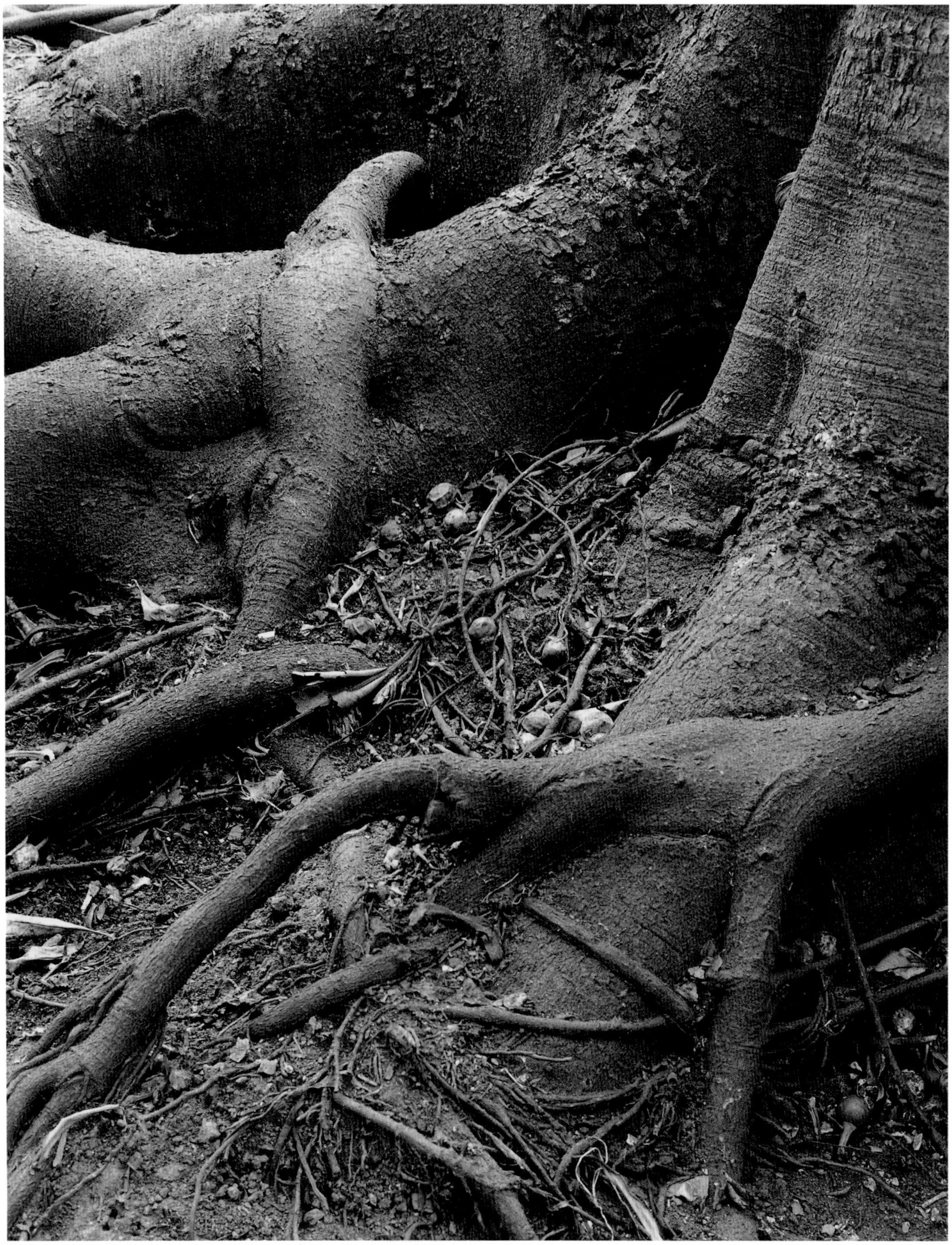

PLATE 92

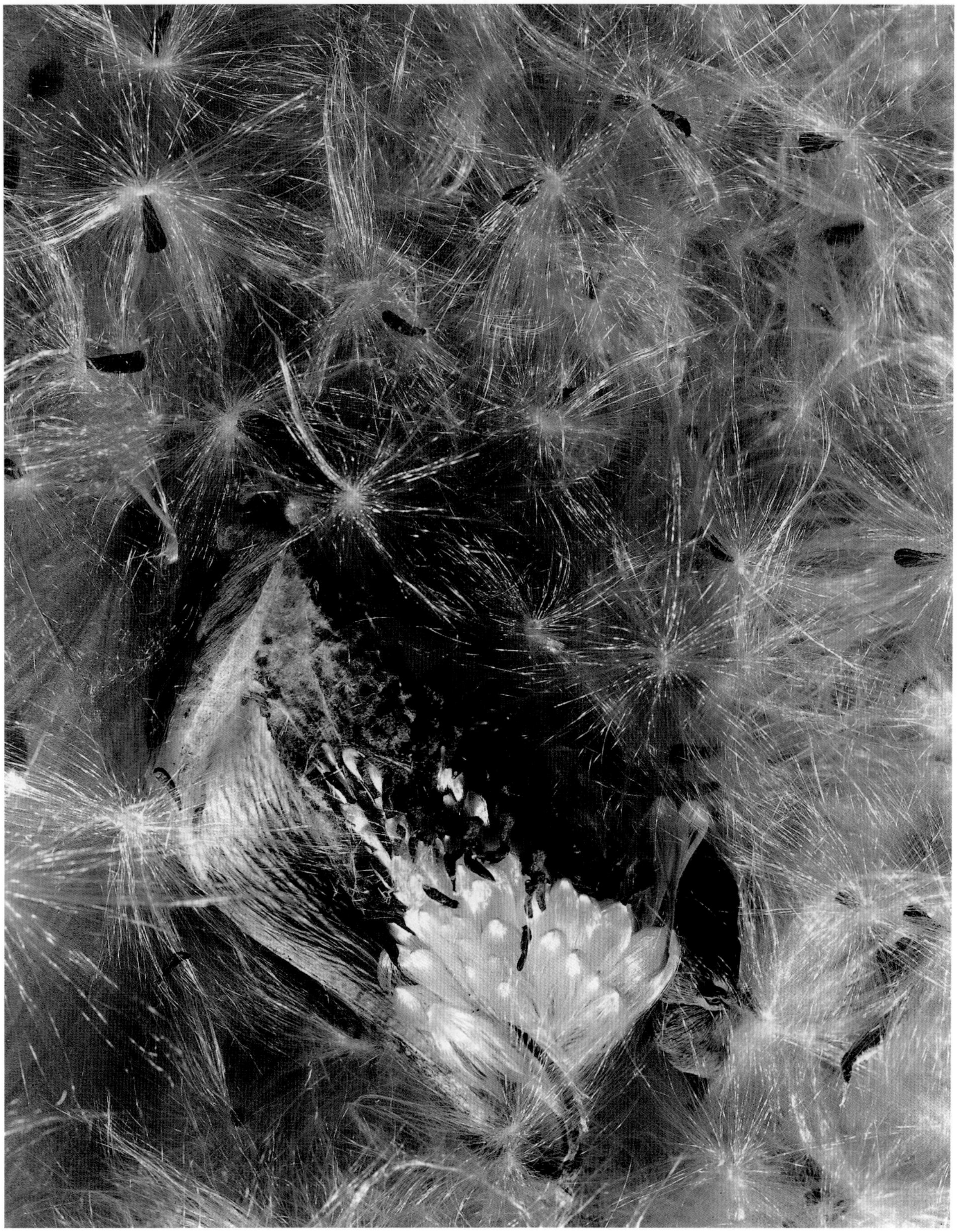

PLATE 93

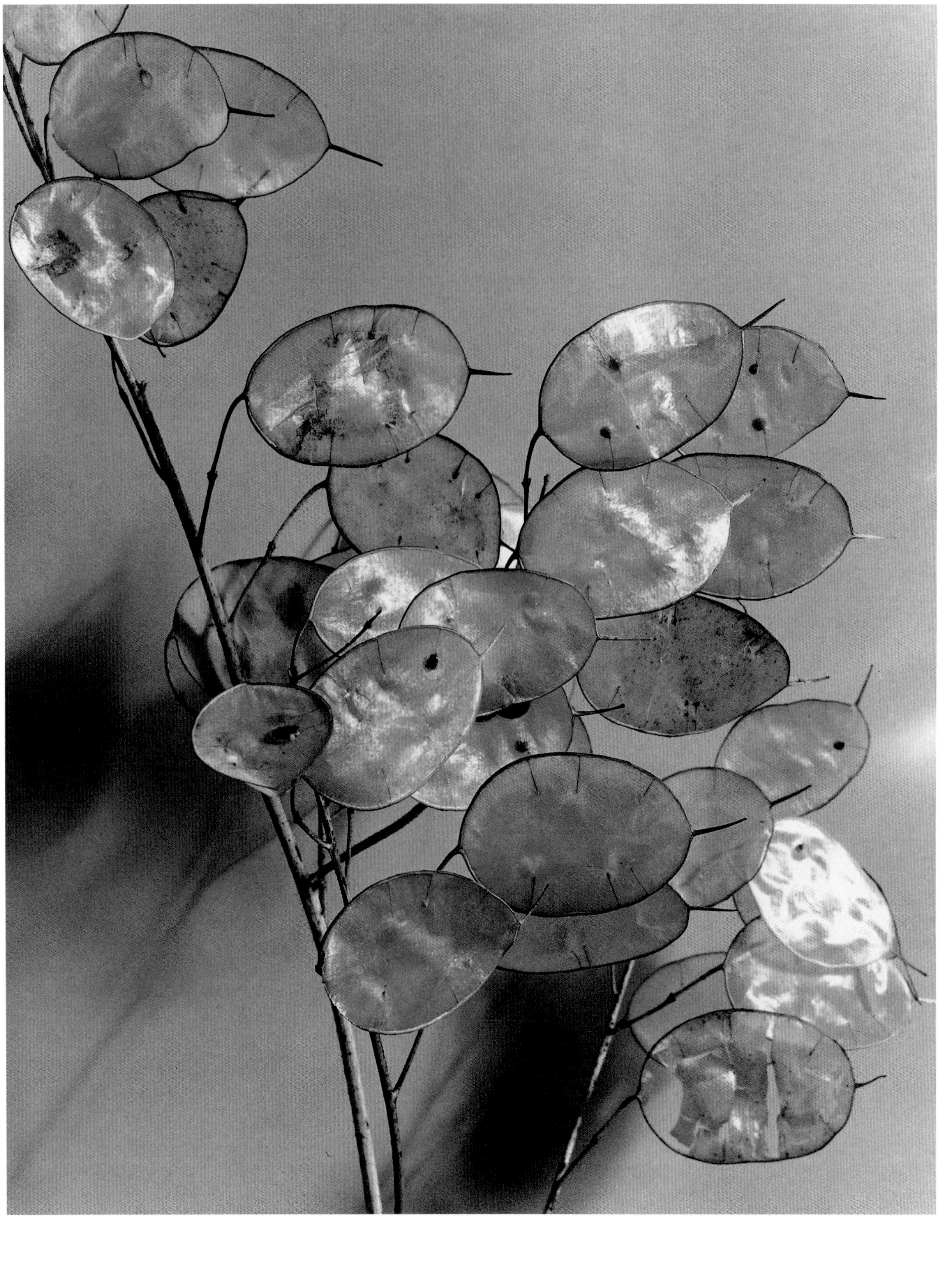

PLATE 94

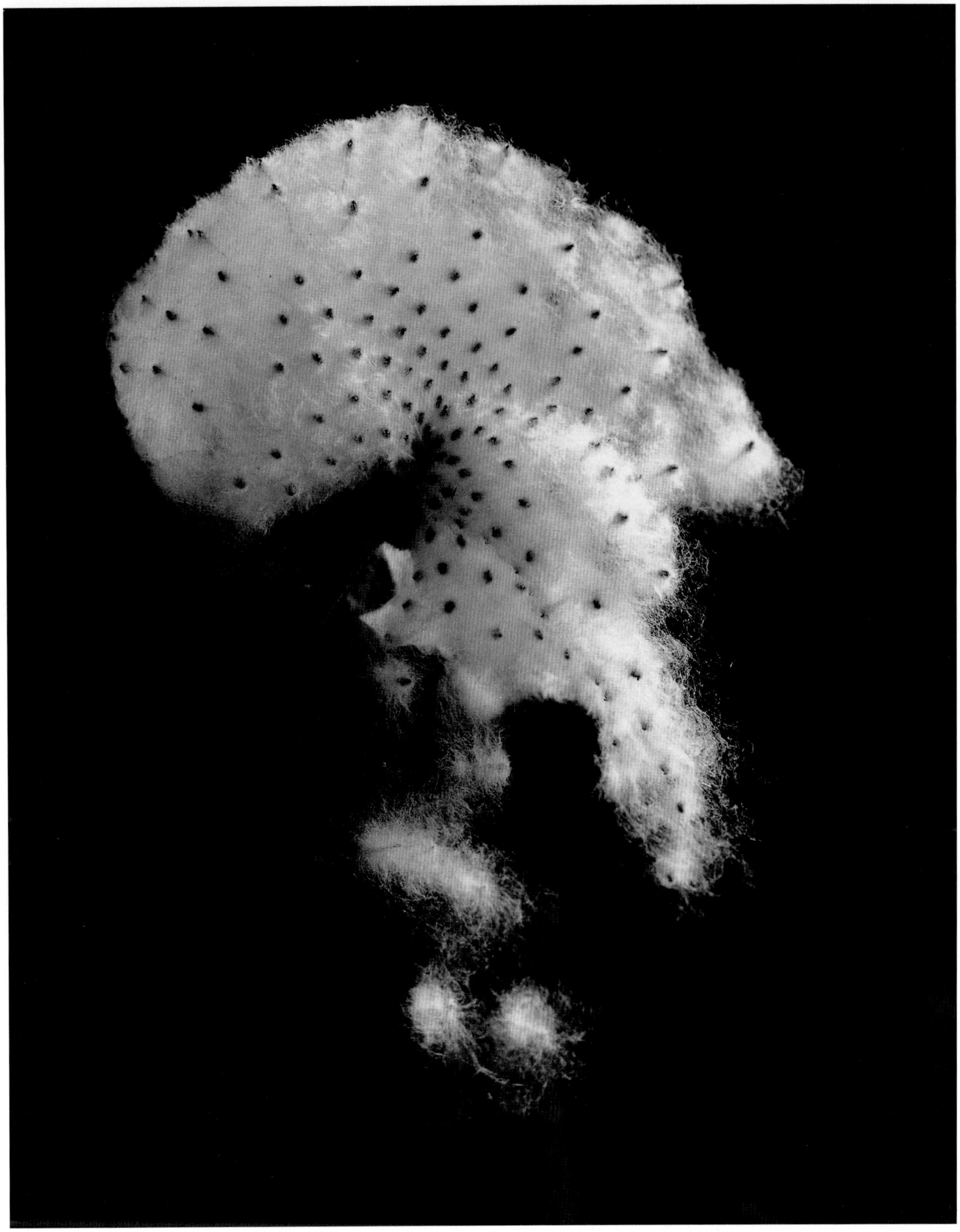

PLATE 95

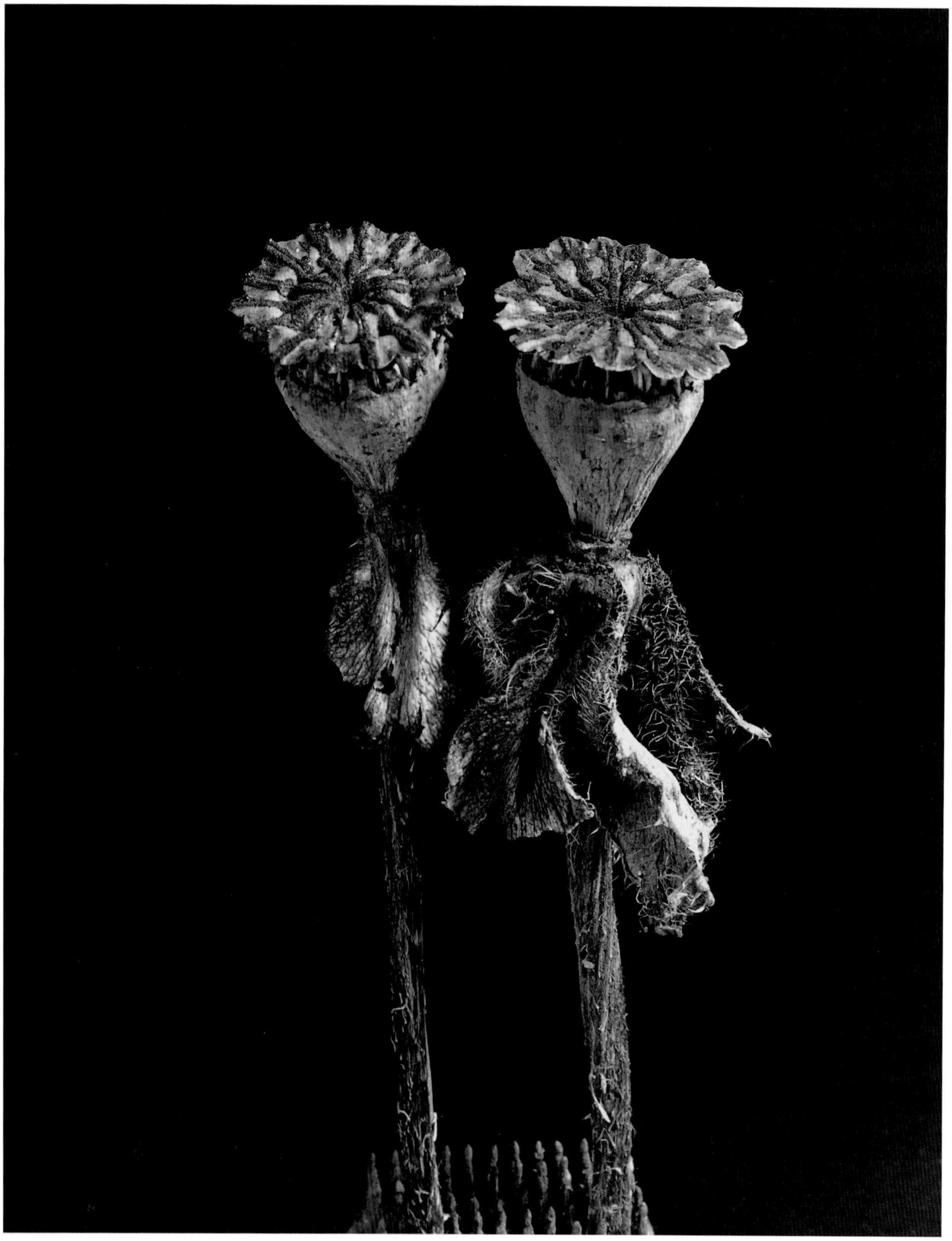

PLATE 96

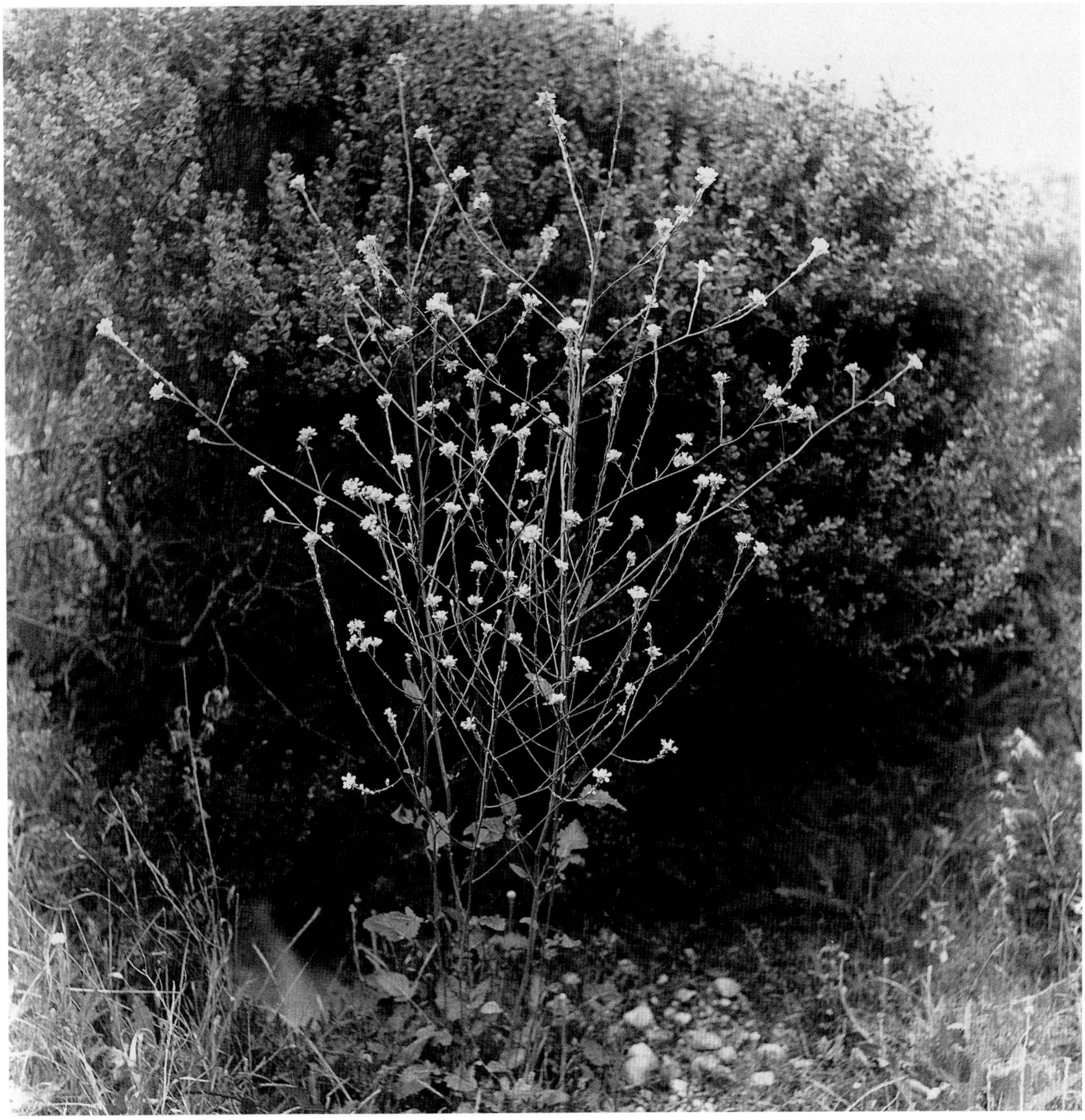

PLATE 97

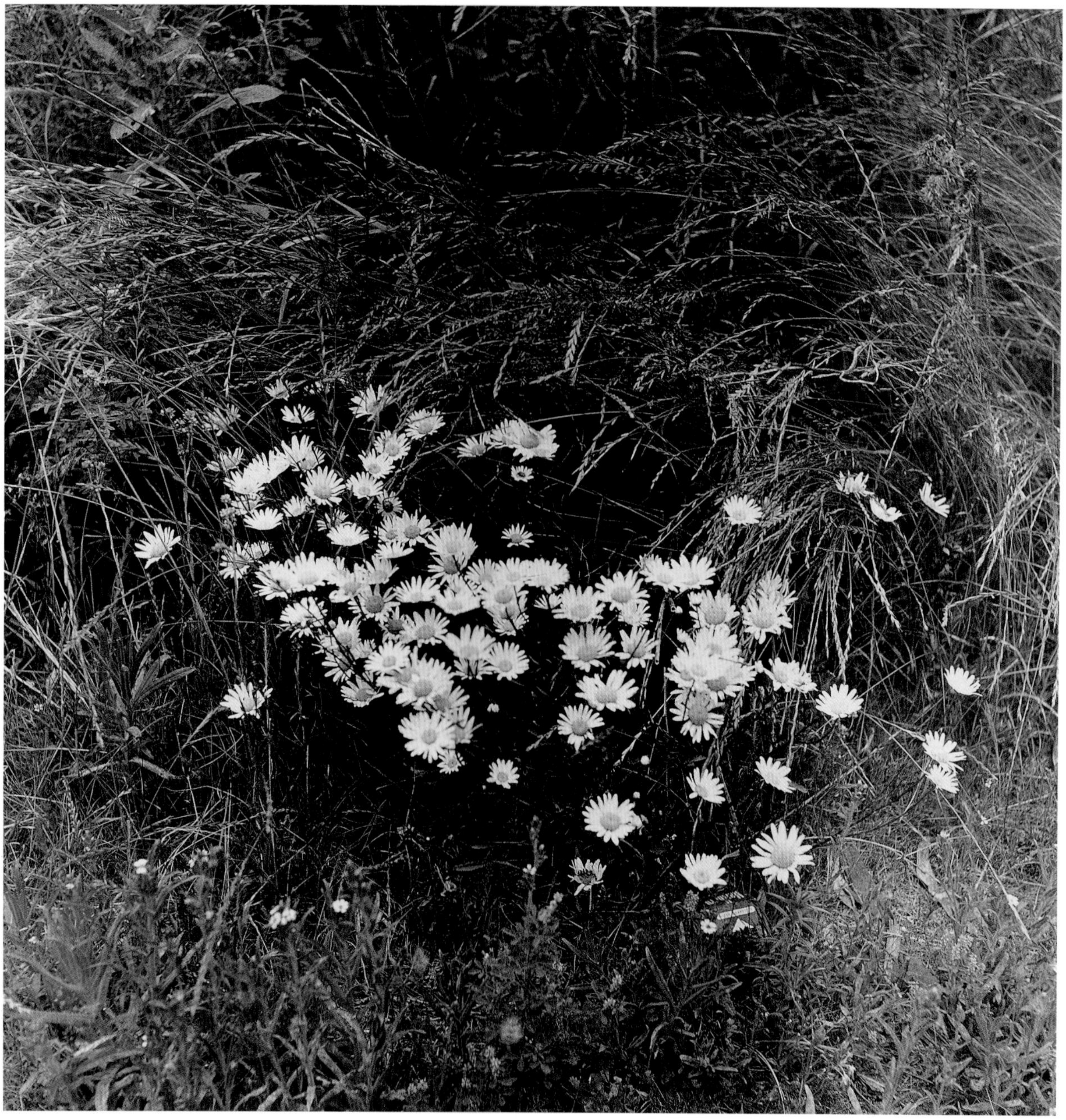

PLATE 98

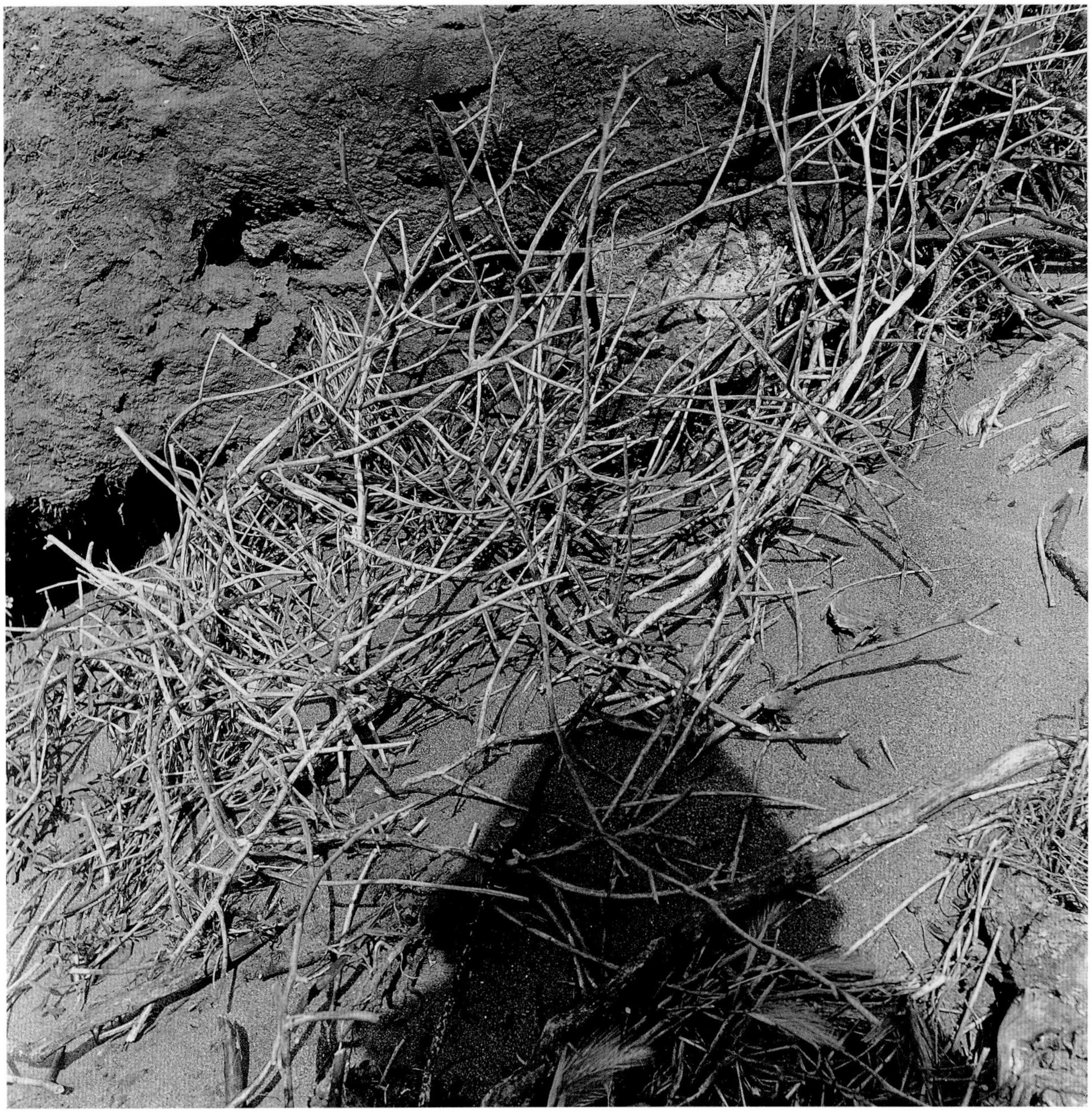

PLATE 99

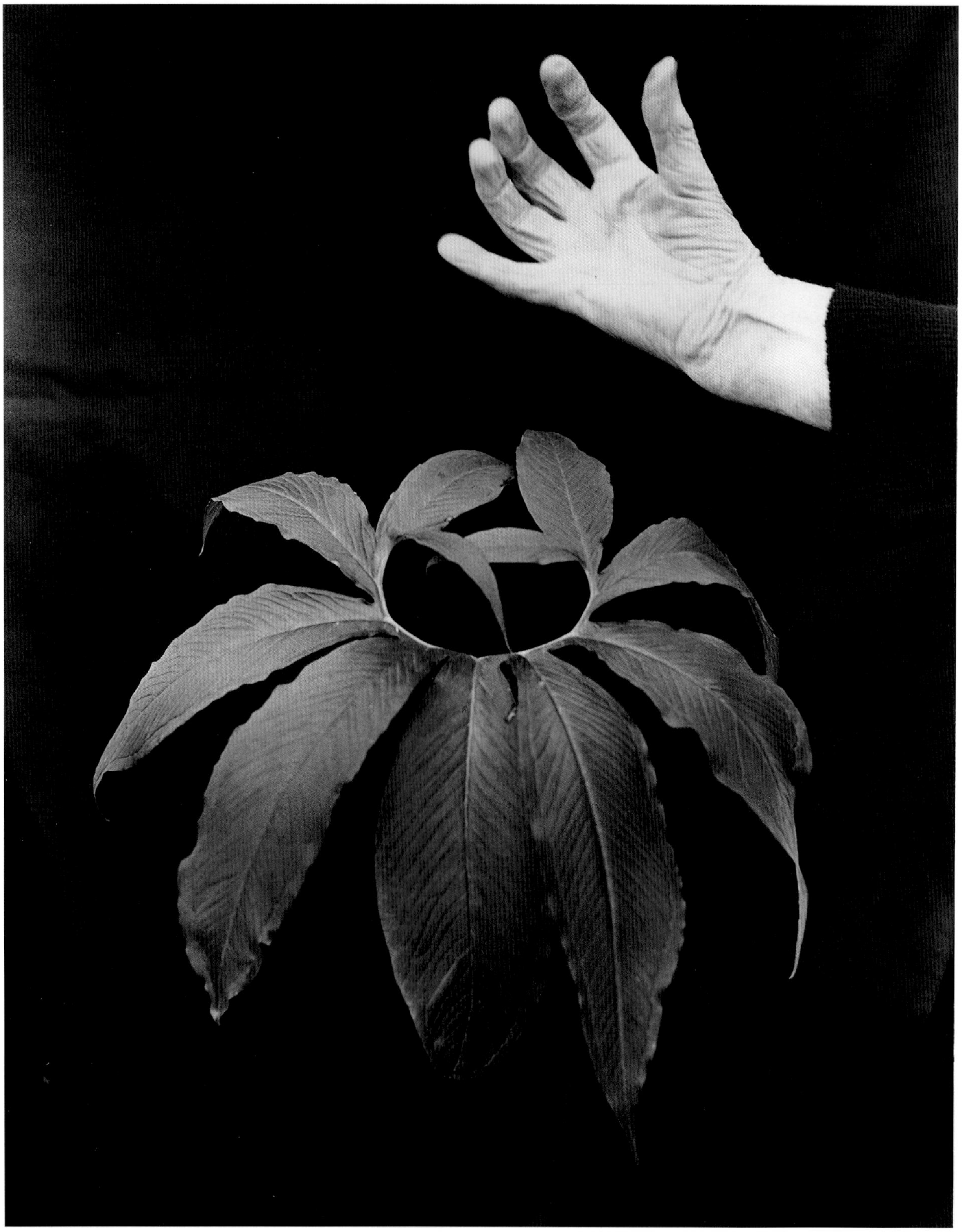

PLATE 100

Notes on the Plates

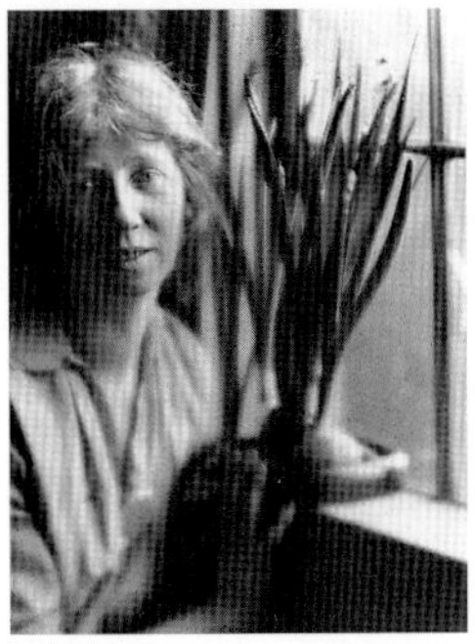

PLATE 1

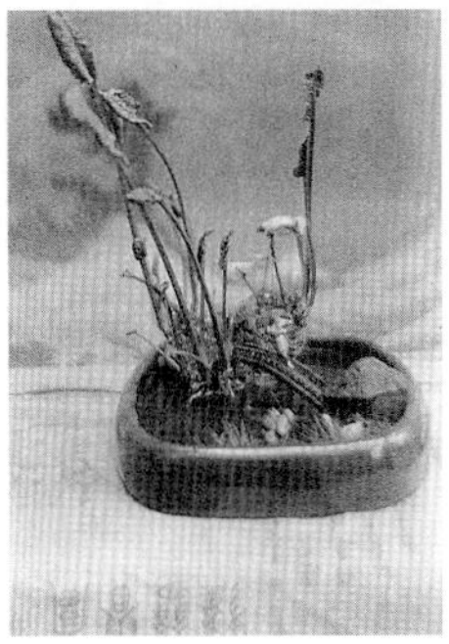

PLATE 2

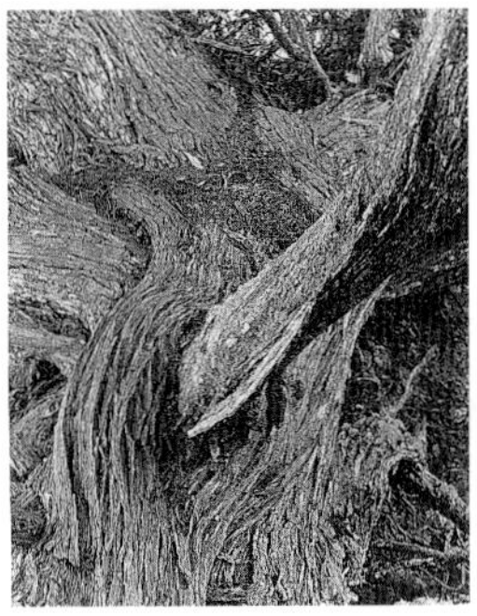

PLATE 3

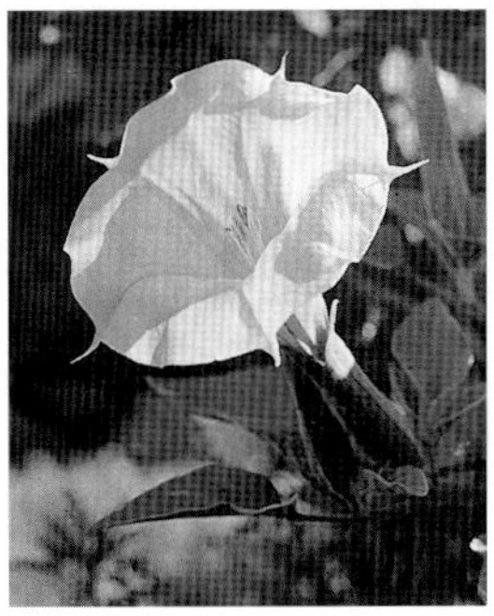

PLATE 4

PLATE 1

Self-portrait, 1913

Narcissus tazetta (variety *papyraceus*)

Derivatives of this species bear clusters of four to eight or more flowers on each stem. The flowers have short coronas and are sweetly fragrant. They appear in a variety of colors. Many types have a white perianth and a yellow corona, but *papyraceus,* the "paper-white," is pure white. The bulbs are often "forced" to bloom indoors on pebbles in a saucer of water. The ancient Egyptians used the flowers for funeral wreaths. Pliny claimed that the flower was called *Narcissus* because of the narcotic quality of its scent; others believe the name refers to Ovid's celebrated youth.

PLATE 2

This Is My Garden, 1910–1915

Pinellia ternata

A native of China and Japan, this tuberous herb has light grayish green, three-parted soft leaves on long stalks. It flowers freely and produces small, light green spathes on slender stems up to 5 inches tall. It is used in Chinese herbal medicine and therapy to reduce expectoration, subdue nausea, strengthen the spleen, stop bleeding, and reduce inflammation.

PLATE 3

At Point Lobos, 1921

Cupressus macrocarpa

The Monterey cypress, a California native, is cultivated in western North America in gardens as hedge and as a windblock in coastal conditions. It grows to 60 feet in the wild but can be twice as tall in cultivation. The tree is narrow and pyramidal-shaped when young, spreading its rich green foliage with age. It flowers in March; the males turn yellow and shed pollen, and the females develop into cones that are 1 to 1½ inches across, made up of four to six scales. The bark is red-brown, ridged, and broken into scales.

PLATE 4

Thorn Apple, about 1921

Datura meteloides

The thorn apple grows about 2 feet high with ovate, toothed, smooth leaves and white funnel-shaped flowers, 4 to 8 inches long. The plant is valued for various medicinal properties. It is used in the treatment of asthma and bronchial complaints and applied externally as an ointment for bruises and swelling. Scopolamine, a pre-anaesthetic used in childbirth and surgery, is found in the seeds, flowers, and foliage. Ingestion of thorn apple in high doses can be dangerous.

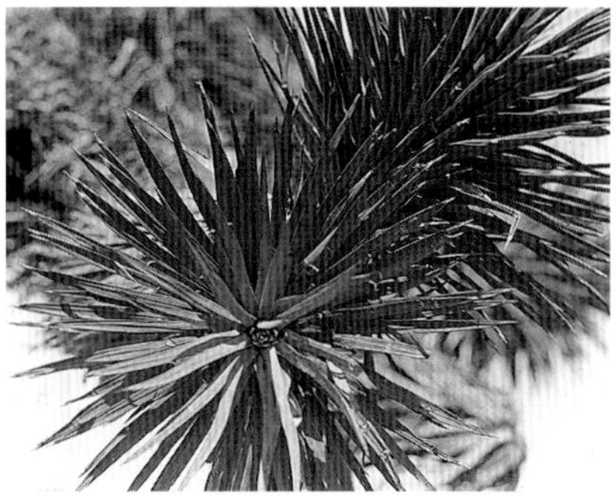

PLATE 5

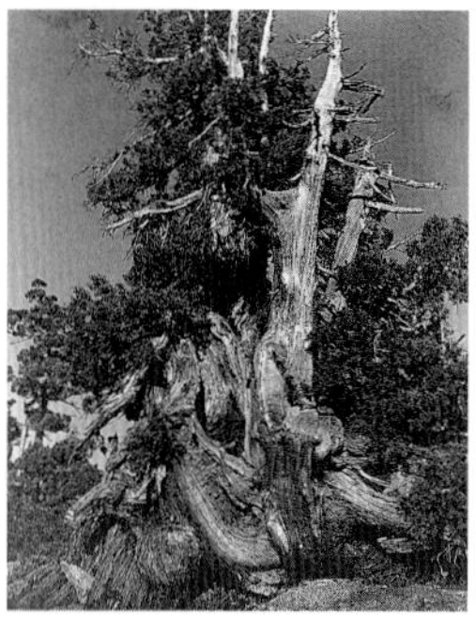

PLATE 6

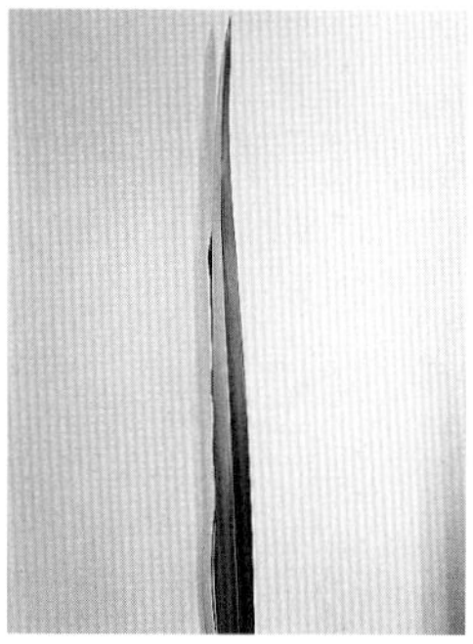

PLATE 7

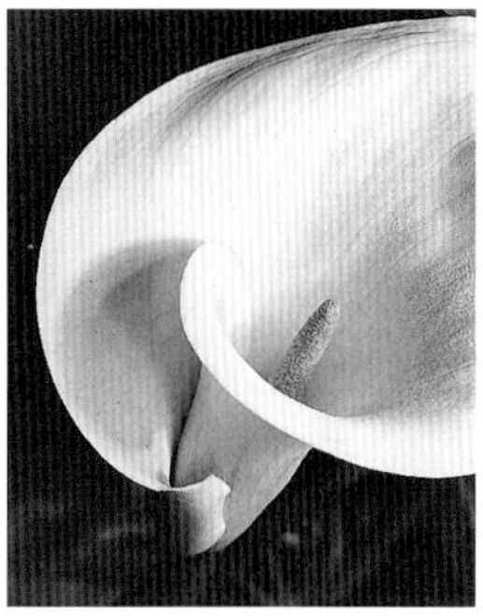

PLATE 8

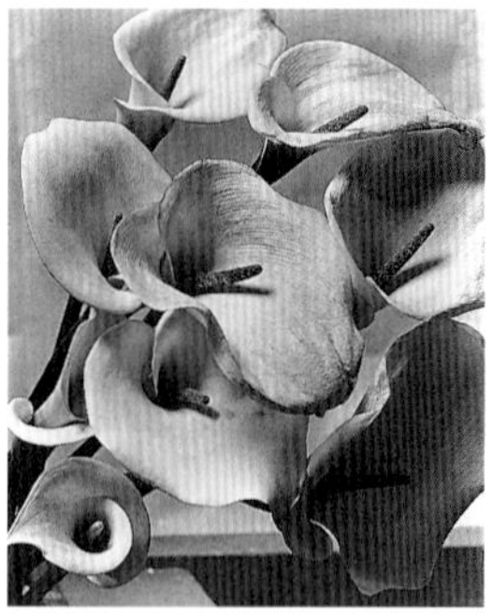

PLATE 9

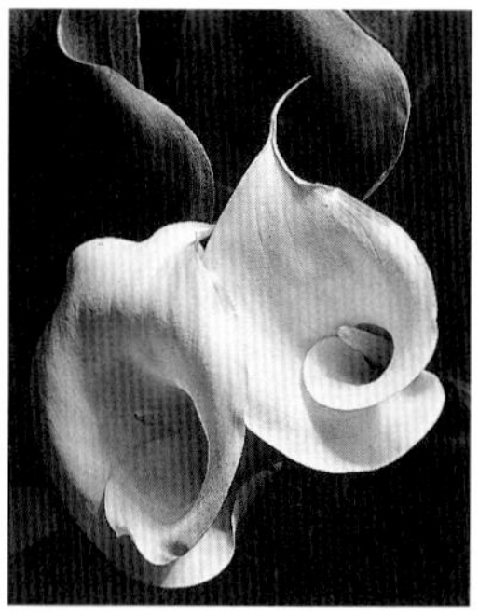

PLATE 10

PLATE 5
Joshua Tree, 1922
Yucca brevifolia
Yuccas are found in desert and semidesert regions of North America and the West Indies, but the Joshua tree primarily grows in areas bordering the Mojave Desert. Up to 30 feet in height, it has stiff, 12- to 20-inch-long, swordlike leaves and forms large panicles of waxy, white, 2-inch, cup-shaped flowers, which open at night and have a sweet fragrance.

PLATE 6
Tree at Donner Pass, 1925
Juniperus occidentalis
The West Coast juniper has a massive trunk with spreading branches. Its diameter can reach more than 20 feet at ground level, although its height does not exceed 80 feet. This evergreen tree has two distinct types of leaves: needlelike on young specimens and mixed needle and scale on mature ones. Its fruits are succulent cones, usually bluish black, and are referred to as berries. The tree has a pungent aroma. The berries of one type of juniper are used to flavor gin, and the wood of several types of junipers is useful as a moth repellent.

PLATE 7
Flax, 1920s
Phormium tenax
The New Zealand flax is a large, dramatic, spreading plant. Its stiff, swordlike, vertical leaves grow in a fan pattern. The leaves can grow up to 9 feet long and 5 inches wide. Their color varies from purple-red to bronze. The plant's long reddish brown flower stalks yield numerous dark red to yellow flowers.

PLATE 8
Calla, about 1925
Zantedeschia aethiopica
The common calla is grown from rhizomes and is a native of South Africa. It forms basal clumps of long-stalked, shiny, rich green, lance-shaped leaves up to 18 inches long and 10 inches wide. The pure white flower bract surrounds a yellow central spike that is covered with tiny true flowers. The flower bracts can be as large as 8 inches long and form on stems up to 3 feet tall. Callas bloom mostly in spring and early summer.

PLATE 9
Callas, about 1925
Zantedeschia aethiopica
See plate 8.

PLATE 10
Two Callas, about 1925
Zantedeschia aethiopica
See plate 8.

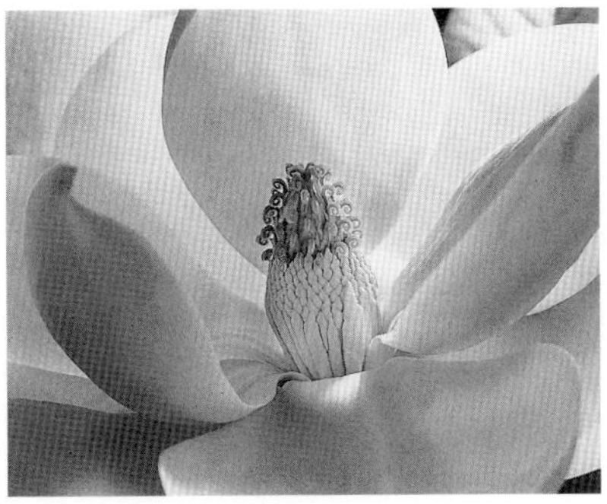

PLATE 11

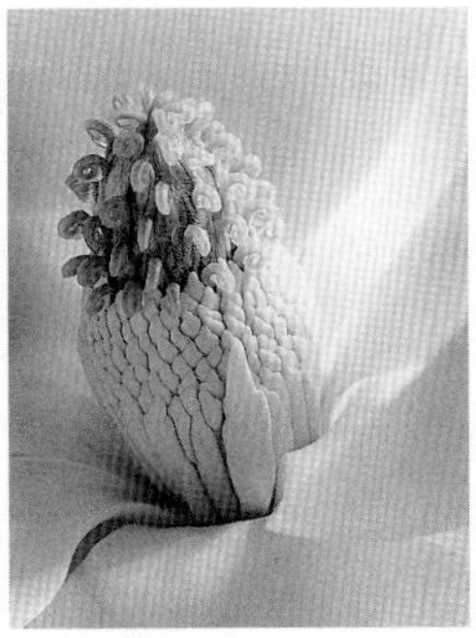

PLATE 12

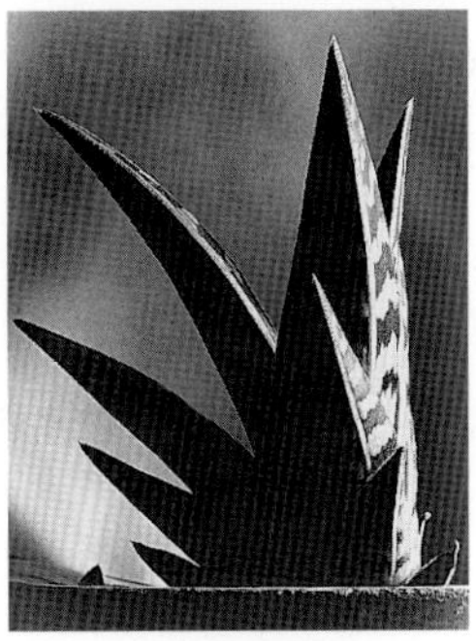

PLATE 13

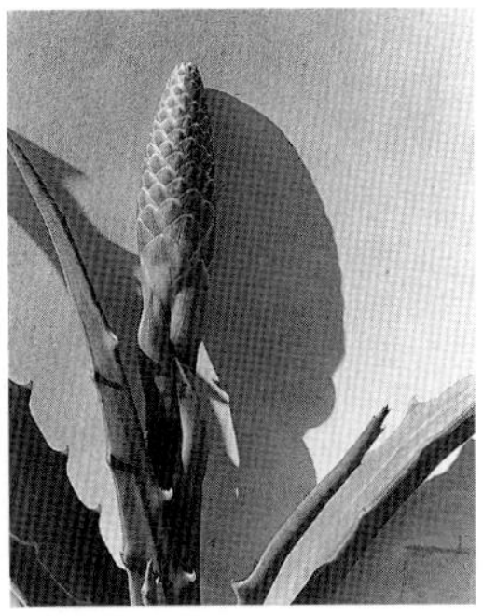

PLATE 14

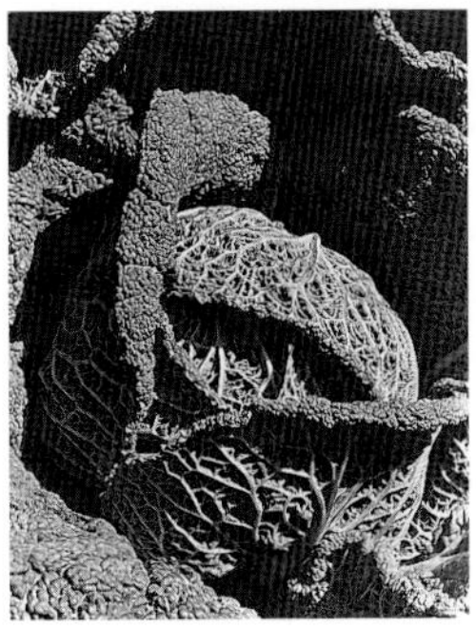

PLATE 15

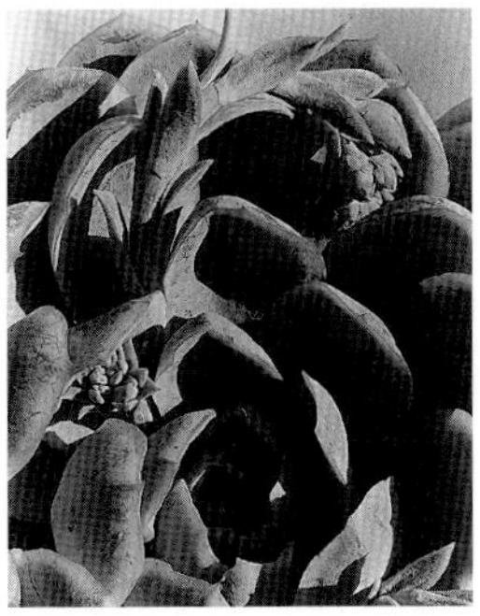

PLATE 16

PLATE 11

Magnolia Blossom, 1925

Magnolia grandiflora

The southern magnolia, or bull bay, is an evergreen tree that grows up to 80 feet high and 40 feet wide. It usually blooms after about fifteen years of age. Its extremely fragrant flowers have six to twelve thick, obovate petals, pure white aging to buff, with purple stamens. The flowers can be up to 8 to 10 inches across and bloom throughout the summer and fall. Each flower lasts from two to four days, and the conelike fruit appears when the petals fall. The leathery leaves are glossy green on top, rusty brown below, and 4 to 8 inches in length. Magnolias are among the oldest plants in the world.

PLATE 12

Magnolia Blossom, Tower of Jewels, 1925

Magnolia grandiflora

See plate 11.

PLATE 13

Aloe, 1925

Aloe variegata

The Partridge breast aloe, or tiger aloe, is a succulent in the lily family. This species grows in a foot-high triangular rosette of fleshy, triangular, 5-inch high leaves which are edged in dark green and banded with white. It forms loose clusters of pink to dull red tubular flowers.

PLATE 14

Aloe, 1920s

Aloe arborescens

The tree aloe, or candelabra aloe, can grow up to 18 feet high. Its branching stems form large clumps of gray-green, spiny-edged, fleshy leaves. During the winter months, it produces bright vermillion to clear yellow flowers in long, spiked clusters.

PLATE 15

Cabbage, 1925

Brassica oleracea

The savoy cabbage is one of the best-known varieties of this species. The outer leaves are tough, heavily textured, and deeply colored. The cabbage is a large terminal bud that is clearly seen when the head is cut in half longitudinally. Cultivated since very early times, cabbages were derived from a wild form native to western and central Europe and western Asia.

PLATE 16

Echeveria 2, 1920s

Echeveria elegans

"Hen and chicks" is a succulent that forms tight rosettes of grayish white, fleshy leaves up to 4 inches wide. Circles of offset young plants surround the mother plant, hence its common name. Its bell-shaped, nodding flowers are pink and lined with yellow and form in 8-inch-long clusters.

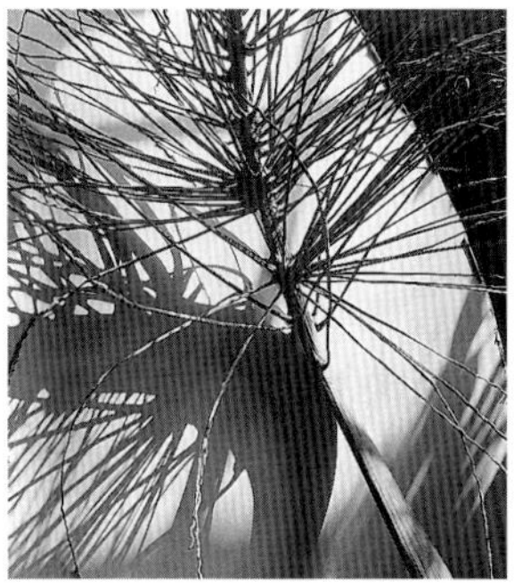

PLATE 17

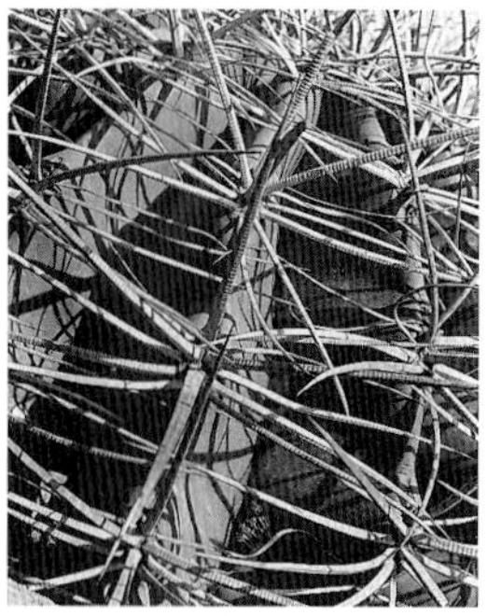

PLATE 18

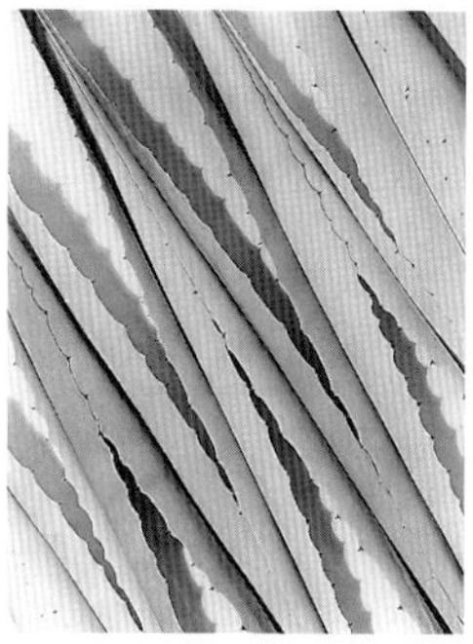

PLATE 19

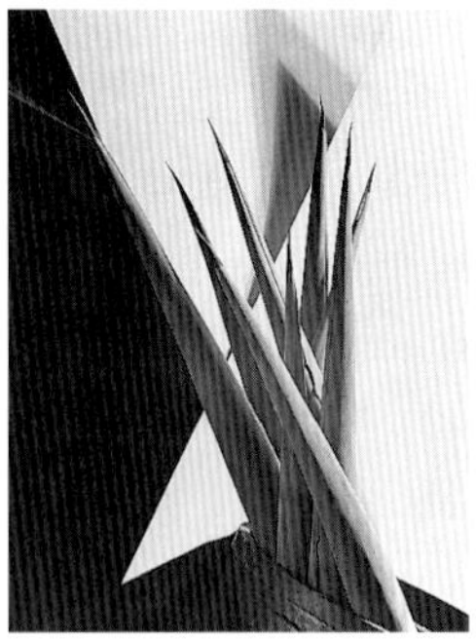

PLATE 20

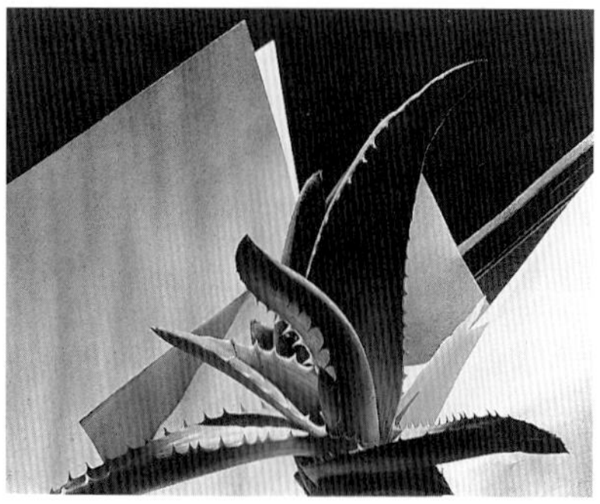

PLATE 21

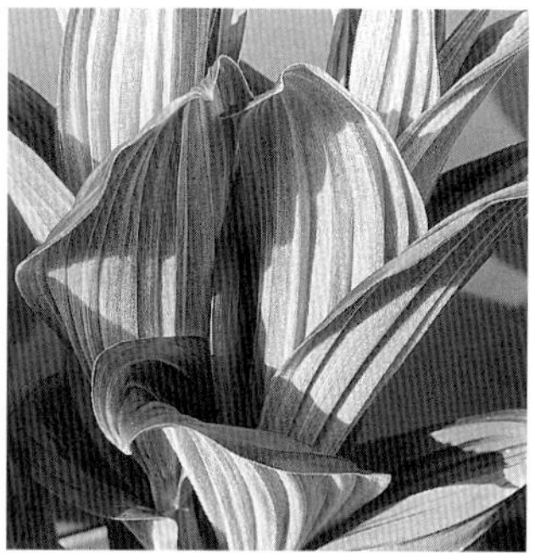

PLATE 22

PLATE 17
Palm Plants, 1925
Phoenix canariensis
Palms are one of the most ancient flowering plants, dating back to the early Triassic period. The Canary Island date palm is a stout palm that grows up to 50 feet high and 40 feet wide. Its trunk can grow to 3 feet in diameter. It has a large crown of stiff, bright green leaves, 15 to 20 feet long. The lower leaflets are spines, which are yellow like the leaf-stalks. Branched flower inflorescences, which are erect to arching, develop from among the foliage. Male and female flowers are usually on separate trees; the females, developing after fertilization, hang in clusters of 1-inch-long oval, yellow to reddish, fleshy fruits.

PLATE 18
Spines, about 1925
Ferocactus (unidentified species)
The barrel cactus is a ribbed and spiny cactus of medium to large size. When young, it has a globular shape, but as it ages, its shape becomes more cylindrical. It flowers during the summer months and produces 3-inch-wide, bell-shaped, yellow, orange, or yellow-edged red flowers.

PLATE 19
Agave Design 2, 1920s
Agave americana
The century plant is a gigantic succulent that forms a large, loose, open, trunkless rosette of spreading broad and thick blue-green leaves up to 6 feet long. The leaves bend downward from the middle and have hooked spines along their edges and at their tips. When about ten or more years old, the plant produces a branched flower stalk 15 to 40 feet high with yellowish green flowers.

PLATE 20
Agave Design 1, 1920s
Agave americana
See plate 19.

PLATE 21
Plant Pattern, 1920s
Aloe (seedling of unknown species)
These succulents in the lily family form clumps of fleshy, pointed, green or gray-green leaves and clusters of cream, yellow, orange, or red flowers. Aloes are good drought-tolerant, ornamental plants.

PLATE 22
False Hellebore, 1926
Veratrum californicum
The stem of the false hellebore, or corn lily, is very stout and leafy and suggests a corn stalk. The plant is 3 to 6 feet high with leaves 6 to 12 inches long. It produces dull white flowers. It is found in meadows in the Sierra Nevada at 4,500 to 8,500 feet, and in the North Coast Ranges and high mountains from Mexico to Washington and east to Colorado.

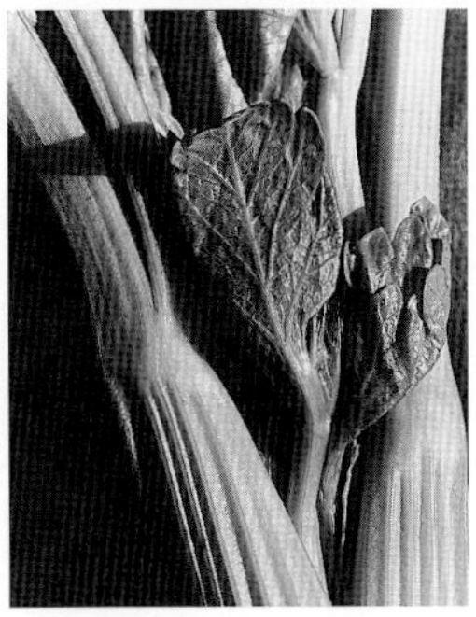
PLATE 23

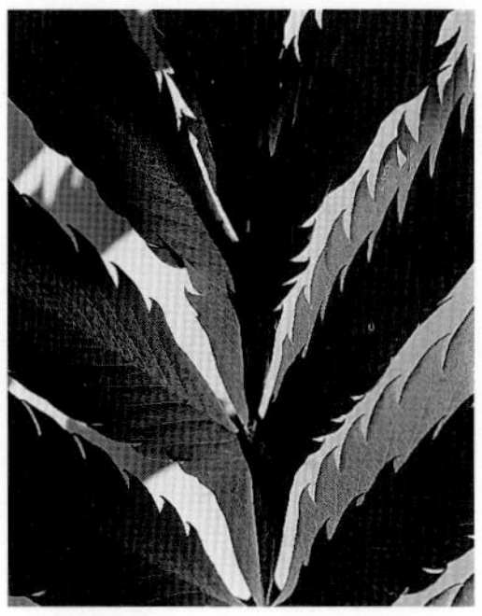
PLATE 24

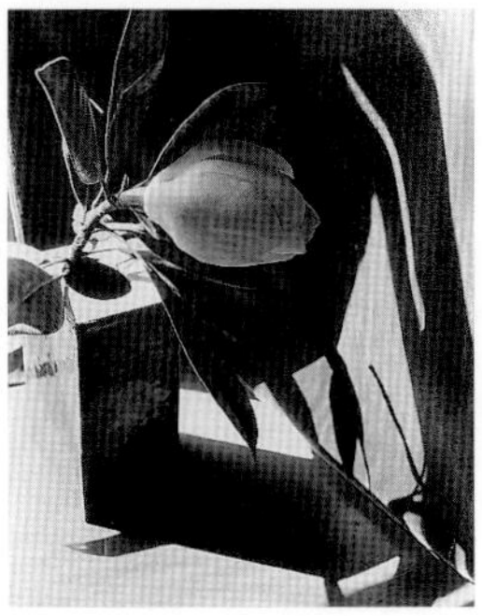
PLATE 25

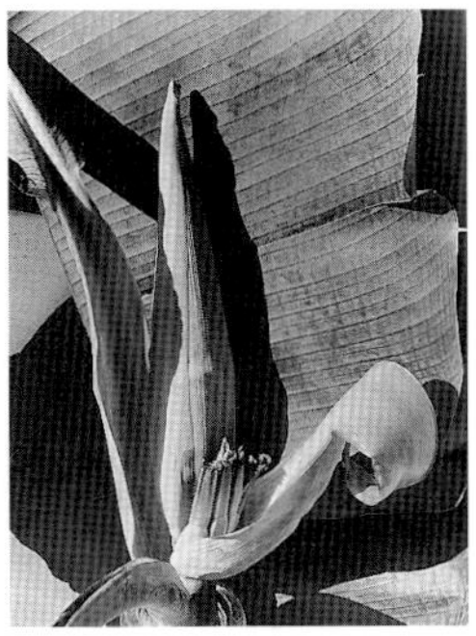
PLATE 26

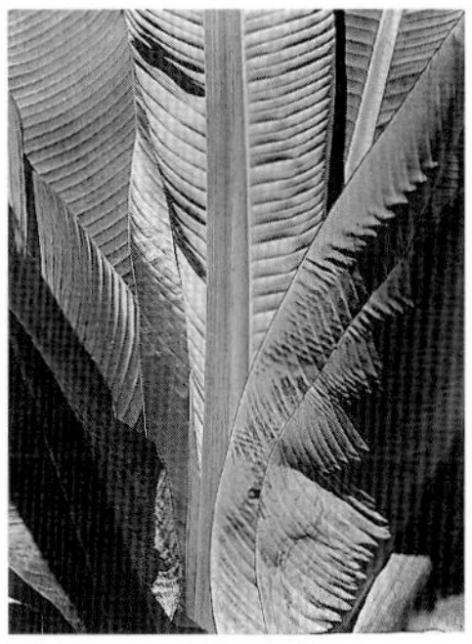
PLATE 27

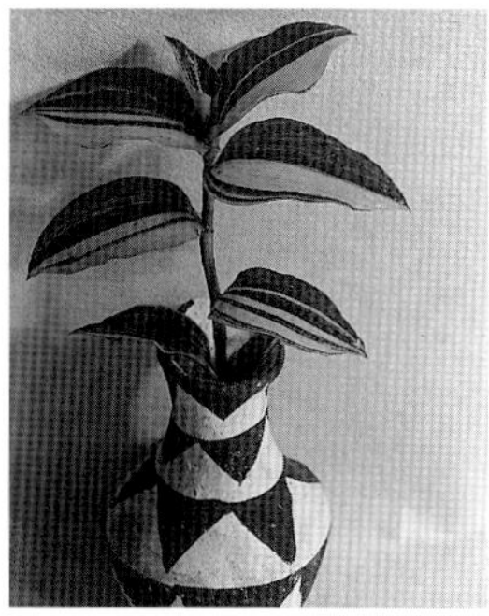
PLATE 28

PLATE 23
Celery 2, about 1925
Apium graveolens
Celery is a biennial grown as an annual that became a table vegetable in the Middle Ages. In ancient Greece, it was considered worthy of crowning the heads of athletes, and, in Plutarch's time, it also was used for funeral wreaths. It has considerable diuretic properties.

PLATE 24
Leaf Pattern, late 1920s
Melianthus major
The honey bush is an evergreen softwood shrub with striking foliage that can grow to 12 to 14 feet high. It can be trained for tall upright stems or as a low spreading bush. The grayish green leaves are up to 12 inches long and are divided into nine to eleven sharply toothed leaflets. When bruised, the leaves emit a disagreeable odor. In late winter to early spring, the plant produces 1-inch-long reddish brown flowers on foot-long spikes.

PLATE 25
Magnolia: Variations on a Theme, 1920s
Magnolia grandiflora
See plate 11.

PLATE 26
Banana Flower 2, late 1920s
Ensete ventricosum
The Abyssinian banana is a palmlike perennial. The lush, arching green leaves with a stout midrib can grow to 10 to 12 feet in length and 2 to 4 feet wide. Within two to five years after planting, flowers form within the cylinder of bronze red bracts at the ends of the stems.

PLATE 27
Banana Plant, late 1920s
Ensete ventricosum
See plate 26.

PLATE 28
Wandering Jew, 1920s
Zebrina pendula
This evergreen perennial houseplant has thick leaves about 1 inch wide and 3 inches long. On top, the leaves are dark olive green with white stripes running lengthwise. The color of the undersides is purple. Occasionally, the plant produces small clusters of purplish-rose and white flowers.

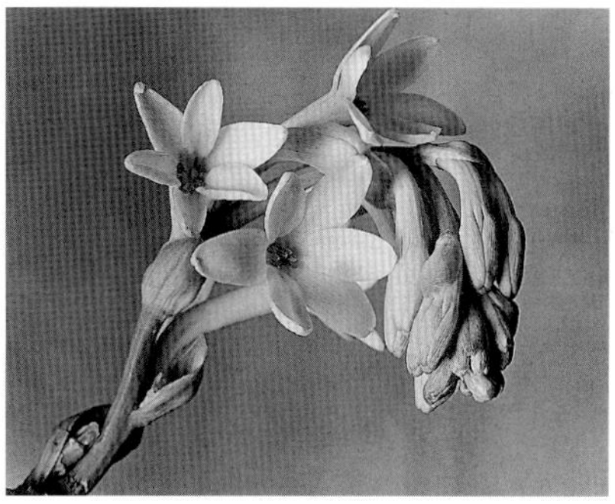

PLATE 29

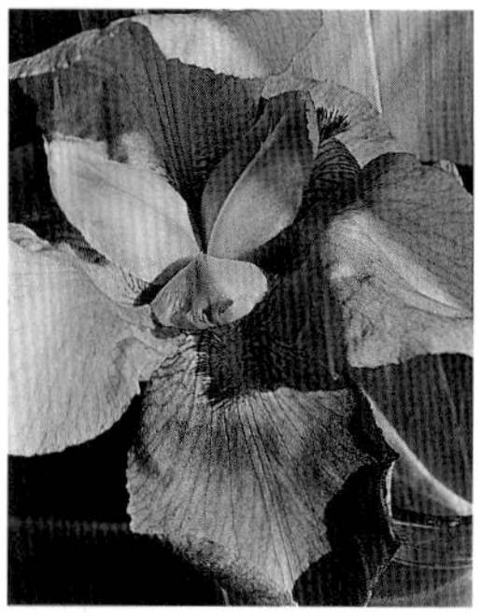

PLATE 30

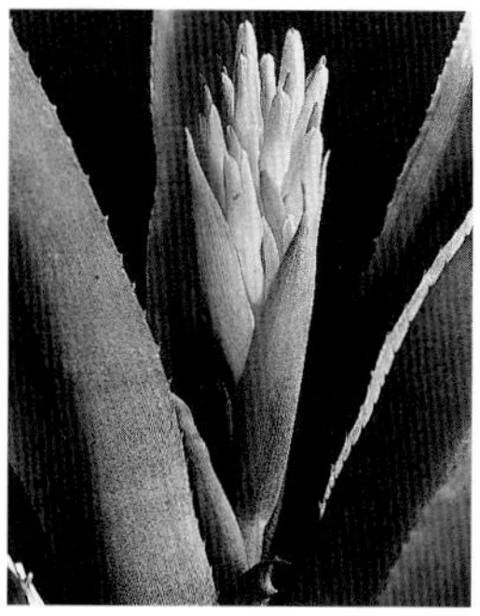

PLATE 31

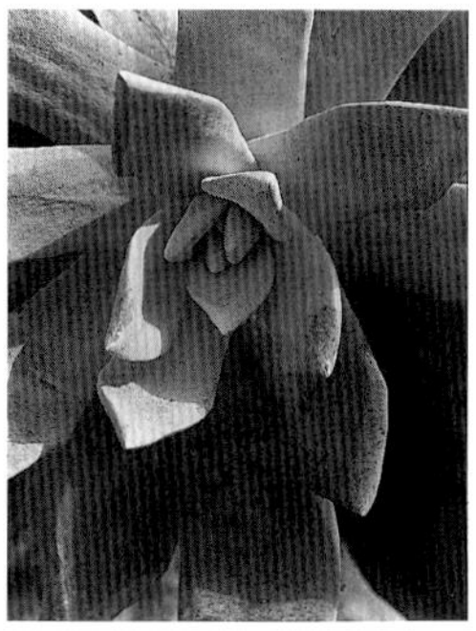

PLATE 32

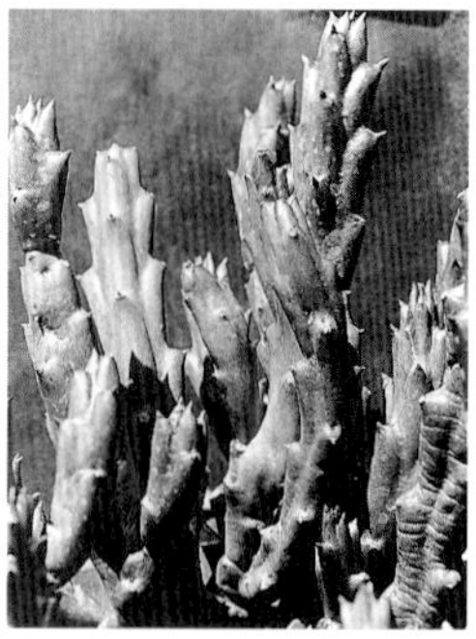

PLATE 33

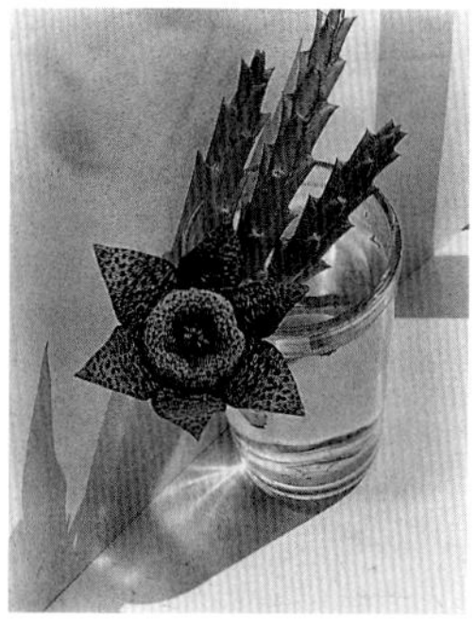

PLATE 34

PLATE 29
Tuberose, 1920s
Polianthes tuberosa
A native of Mexico, the tuberose is known for its powerful, sweet fragrance. The plant blooms from summer to fall and produces flowers that are white, tubular, and loosely arranged in spikelike clusters on up to 3-foot-long stems. The long, narrow, grasslike leaves form around the base.

PLATE 30
Iris, 1920s
Iris germanica
Grown from rhizomatous rootstock, bearded irises are complex structures. There are three inner erect standards surrounded by three outer petal-like falls on which are found the characteristic tufts of hairs. The flowers bloom in mid-spring on stems that grow 2½ to 4 feet high. The leaves are swordlike, up to 18 inches long, and overlap each other to form a flat fan of foliage.

PLATE 31
Billbergia, 1925
Billbergia vexillaria
This evergreen perennial is a native of Brazil and is related to the pineapple. It has 1- to 2-foot-long stiff, spiny-toothed leaves in basal clusters out of which rises a taller, erect flower spike. The tubular flowers are purple and the lower bracts are long, pointed, and red.

PLATE 32
Ghost Plant, 1920s
Dudleya farinosa
Dudleya is a rosette-forming succulent native to California, Arizona, and Mexico. The leaves are fleshy and covered with a heavy coating of chalky powder. The 4-inch-diameter plant produces flowers in white, pale yellow, yellowish pink, and bright red. This species is found along the Monterey coast and forms clusters of small white rosettes.

PLATE 33
Stapelia 2, 1928
Orbea variegata
The starfish flower plant is a succulent that resembles a cactus. It forms in clumps of four-sided, spineless stems up to 6 inches long. The fleshy five-pointed flowers are 3 inches wide, yellow with dark purplish brown markings, and accented by a prominent inner ring. The flowers exude a carrion smell to attract flies for the transfer of pollen.

PLATE 34
Stapelia in Glass, 1928
Orbea variegata
See plate 33.

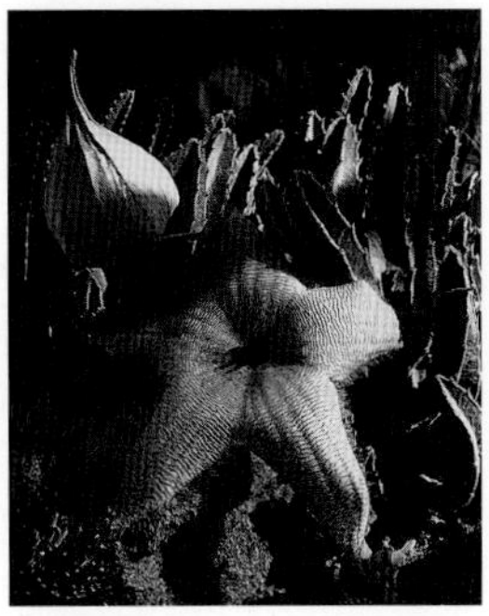
PLATE 35

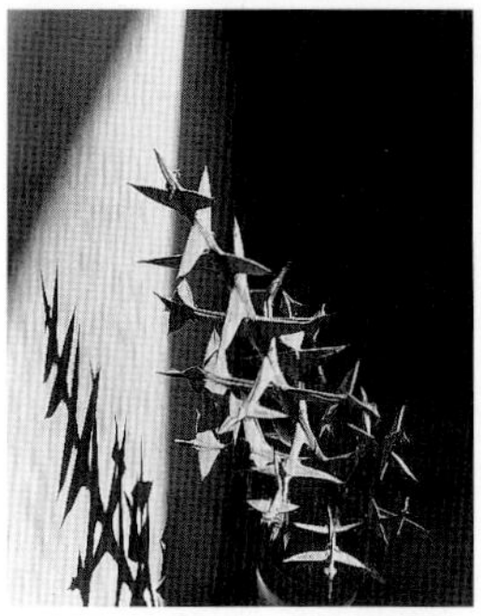
PLATE 36

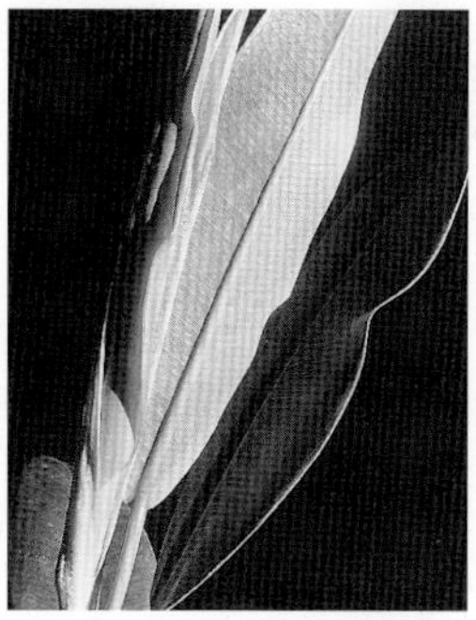
PLATE 37

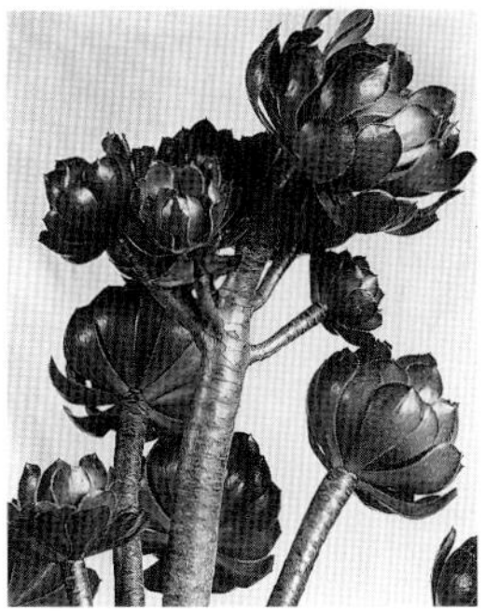
PLATE 38

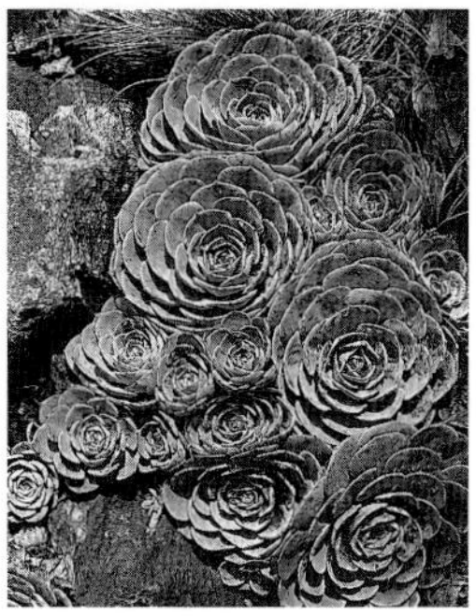
PLATE 39

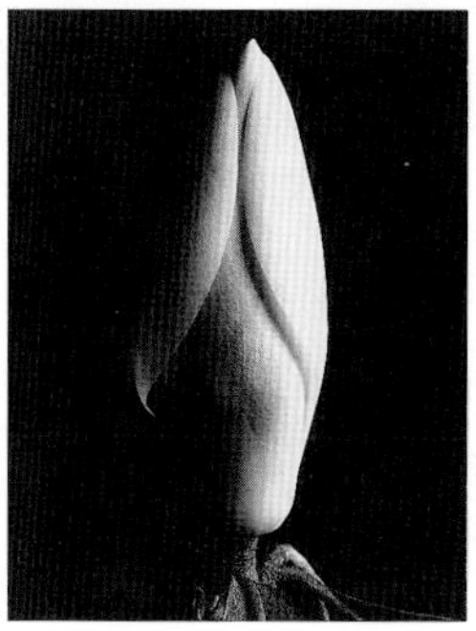
PLATE 40

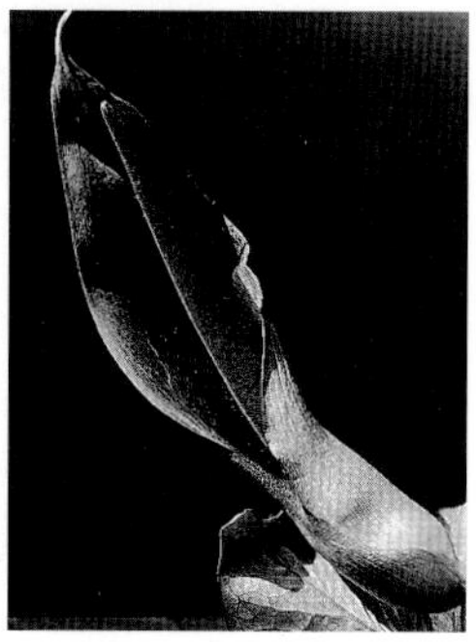
PLATE 41

PLATE 35
Stapelia Flower, 1928
Stapelia gigantea
This odd, fleshy, cactuslike plant produces large showy flowers. The dull light green stems are four-sided and grow up to 9 inches tall. The 10- to 16-inch-wide, five-pointed starlike flowers have shallowly depressed centers and fringed edges. They are pale yellow in color with transverse purplish brown markings and are covered with purplish or crimson hairs. The flowers have an offensive odor.

PLATE 36
Colletia Cruciata 7, 1929
Colletia cruciata
The anchor plant is a curious, rigid, glossy green, spiny shrub with flattened branches armed with flat triangular spines. These sharply pointed spines are actually modified branches borne in pairs, each set at right angles to its neighbor. The plant has tiny deciduous leaves, and, in autumn, it forms small, yellowish white tubular flowers.

PLATE 37
Rubber Plant, late 1920s
Ficus elastica
The rubber plant, a native of India and Malaysia, is an evergreen shrub or tree that reaches up to 40 feet high. Its dark green, glossy leaves are thick and leathery, 8 to 12 inches long and 4 to 6 inches wide. New leaves unfold from a pink sheath that withers and drops. The rubber plant can be grown outdoors in areas of California, but it is generally a houseplant.

PLATE 38
Aeonium, 1920s
Aeonium arboreum (variety *atropurpureum*)
This decorative succulent forms thick, erect, branched stems up to 3 feet tall. Each branch develops a striking 6- to 8-inch-wide rosette of dark purple, lightly fringed, fleshy leaves. It produces golden yellow flowers.

PLATE 39
Hen and Chickens, 1929
Echeveria elegans
See plate 16.

PLATE 40
Magnolia Bud, 1920s
Magnolia grandiflora
See plate 11.

PLATE 41
Black Lily, 1920s
Arum palaestinum
The black calla is a perennial with tuberous roots and arrow- or heart-shaped leaves, 6 to 8 inches long. The curious callalike blossoms are on short stalks. The arrow-shaped flower bract, about the same length as the leaves, is green outside and blackish purple within the curved back, half enclosing the blackish purple spike. In California, black callas bloom in spring and early summer.

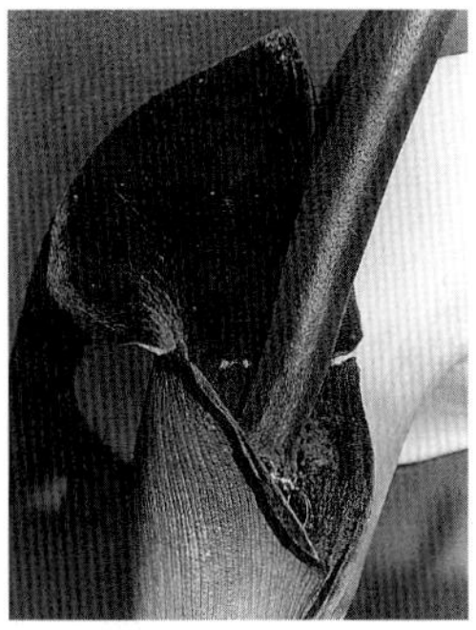

PLATE 42

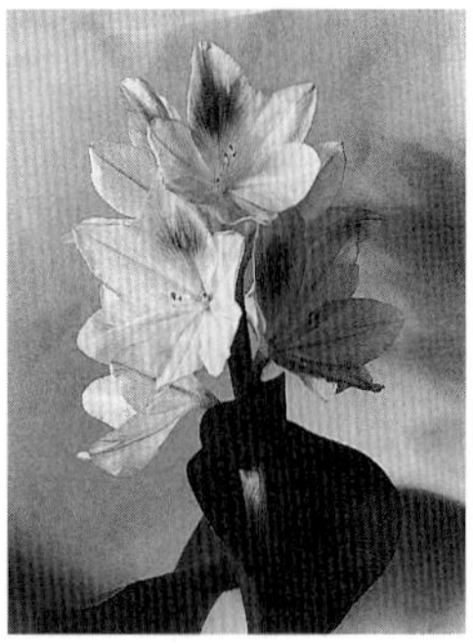

PLATE 43

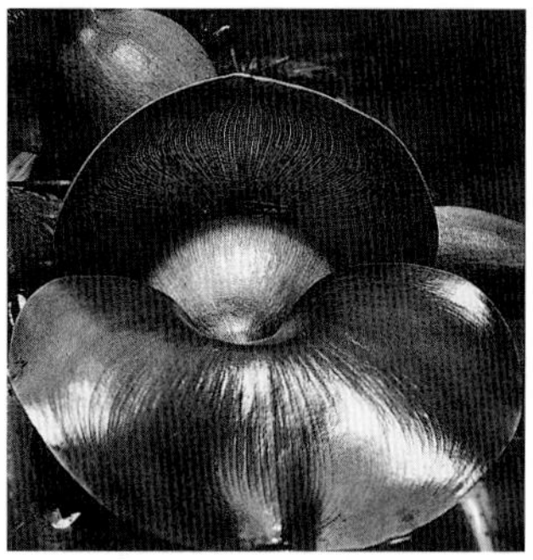

PLATE 44

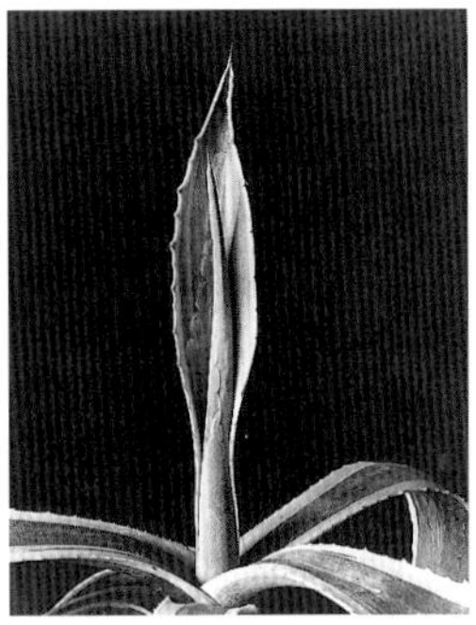

PLATE 45

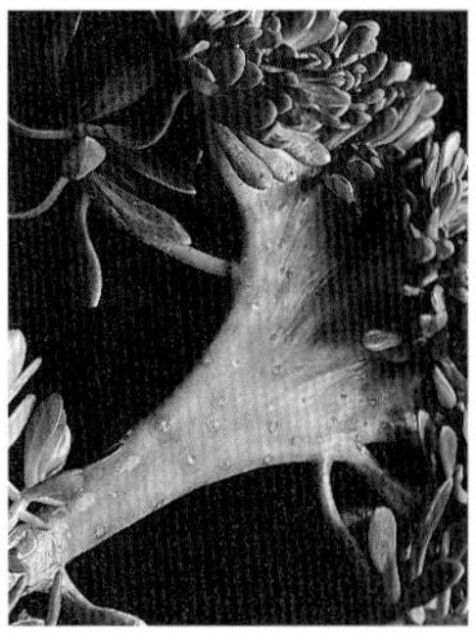

PLATE 46

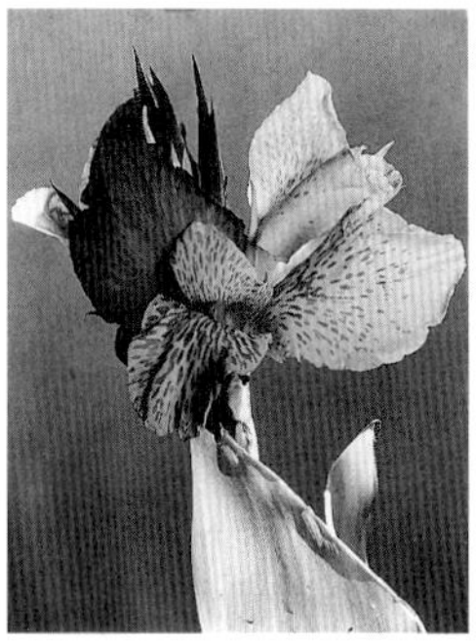

PLATE 47

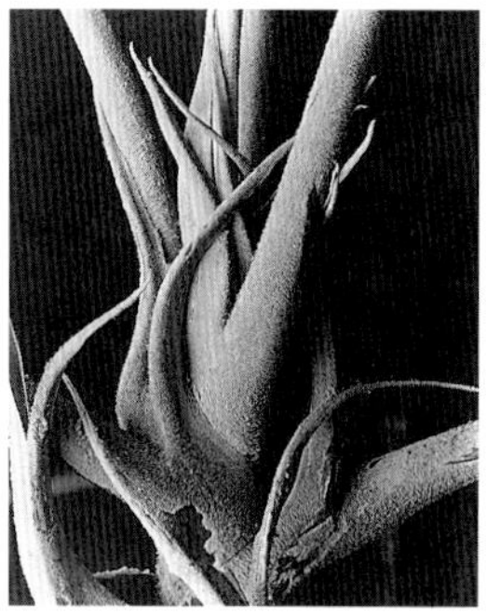

PLATE 48

PLATE 42
Black and White Lilies 3, 1920s
Arum palaestinum and *Zantedeschia aethiopica*
See plates 41 and 8.

PLATE 43
Blossom of Water Hyacinth 2, 1920s
Eichhornia crassipes
The water hyacinth is an aquatic plant native to tropical America. It has floating leaves and feathery roots. The leaves are ½ to 5 inches wide and nearly circular in shape with the leaf stems inflated. The 2-inch-wide blooms are arranged on many-flowered spikes and are lilac blue with a yellow spot in the center of the upper petals.

PLATE 44
Water Hyacinth 1, 1920s
Eichhornia crassipes
See plate 43.

PLATE 45
Agave, 1920s
Agave americana variegata
See plate 19.

PLATE 46
Sedum Cristate, 1920s
Sedum praealtum cristatum
Green cockscomb, an evergreen tree sedum, is common and hardy in warm temperate areas. It is a light green succulent that grows into a dense bush up to 3 feet high with fleshy, shiny, flat as well as round stems and spatulate leaves. When it blooms, its yellow flowers have long narrow petals. The cristate form is bizarre and an artist's delight. The plant was grown in Central America for its medicinal properties.

PLATE 47
Canna 2, 1920s
Canna generalis 'Cleopatra'
A tropical native, this canna is grown from tuberous rootstock and forms stalks 3 to 6 feet tall. The large leaves are green to bronze red in color. The 8- to 10-inch-long yellow flowers are strikingly streaked and spotted with red.

PLATE 48
Fatsia Papyrifera, 1930
Fatsia papyrifera
This hardy shrub or small tree is usually 5 to 7 feet high, but outdoors in southern California it can grow to 20 feet. Its branches and young leaves are covered with down. Mature leaves reach 1 foot in length and have five to seven lobes. Rice paper is produced from this plant's white pith.

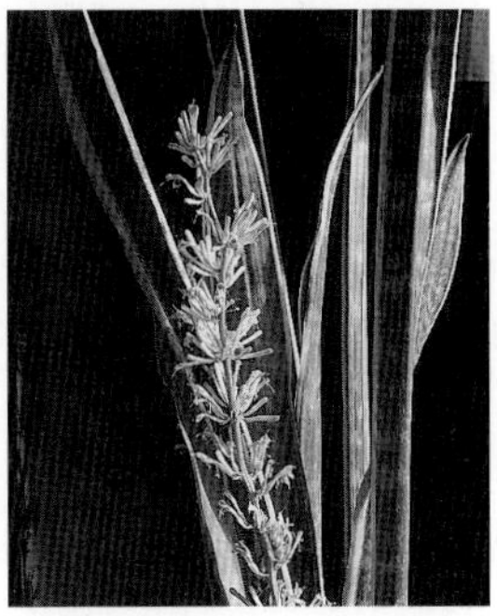

PLATE 49

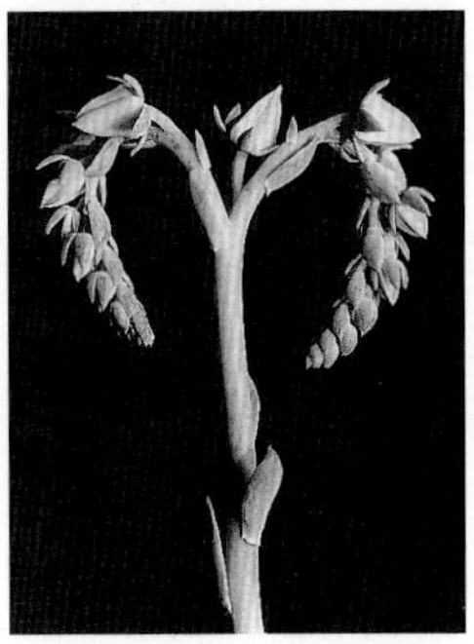

PLATE 50

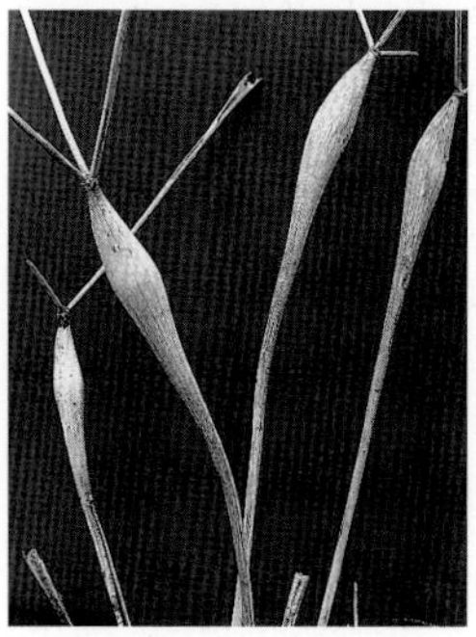

PLATE 51

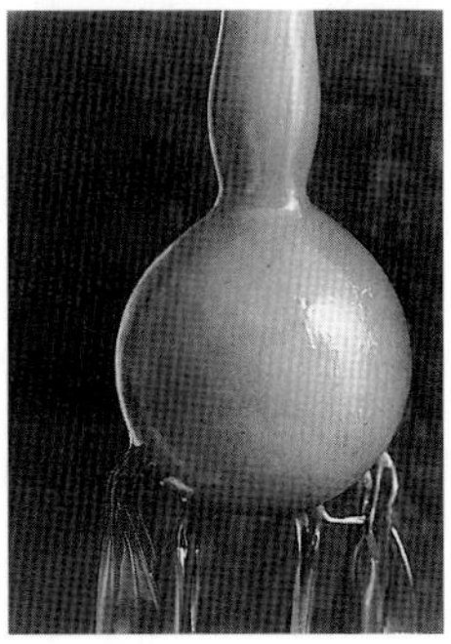

PLATE 52

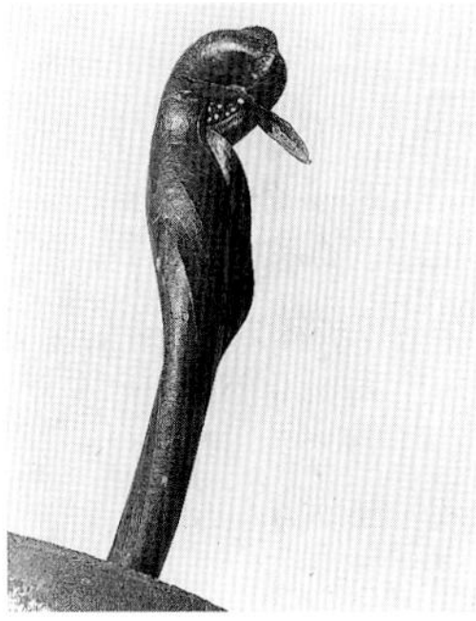

PLATE 53

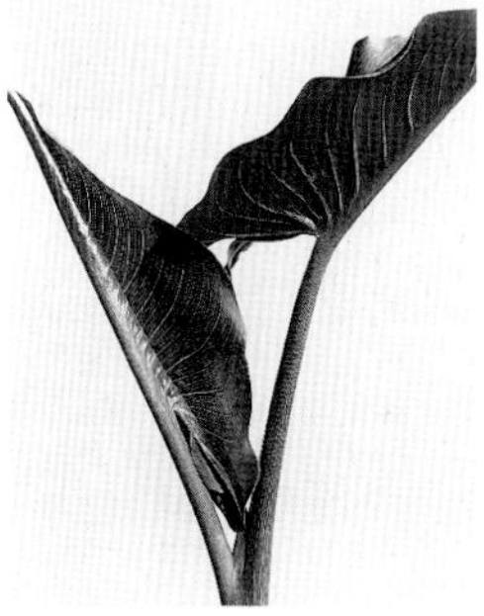

PLATE 54

PLATE 49
Mother-in-law's Tongue, about 1930
Sansevieria trifasciata
Also called snake plant, this evergreen perennial can be grown outdoors or as a houseplant. The toughness of its leaves and ability to survive frequent neglect account for its common name, mother-in-law's tongue. Its leaves grow in rosettes and are up to 4 feet tall and 2 inches wide. They are thick, rigid, and upright, and their dark green patterning with gray-green banding resembles certain snakeskins and accounts for the plant's other common name.

PLATE 50
Echeveria, about 1930
Echeveria elegans
See plate 16.

PLATE 51
Wild Buckwheat, late 1920s
Eriogonum inflatum
Commonly called desert trumpet, or Indian pipe weed, this ½- to 3-foot-tall plant is an annual or perennial herb or small shrub with basal, roundish leaves about 1½ inches long. Several stems form from the base and each develops a diffused panicle. The plant is native to most areas of the West. Flower clusters are often used in dried arrangements.

PLATE 52
Kelp, about 1930
Nereocystis luetkeana
Bull kelp, or sea nymph bladder, is the most massive kelp of California. From deep water, it grows as a long, hollow stipe up to 115 feet, gradually enlarging at the top to a huge, elongate air chamber and terminal bulb. The bulb supports thirty-two to sixty-four blades, each up to 11 feet long, which hang down into the water. The blades and stipe of bull kelp were used by Pacific Coast Indians in a variety of food dishes.

PLATE 53
Pitcher Plant, about 1930
Darlingtonia californica
The California pitcher plant, also called calf's head, is a greenish yellow plant, 8 to 15 inches high, with leaves enlarged upward into a rounded hood. One side of this hood has a circular orifice covered by a two-forked appendage. The cavity in the leaves contains a liquid with properties similar to gastric juice. Flowers are borne on a scape and have yellowish green sepals with purple stripes and dark purple petals. The plant grows in marshy meadows and along streams.

PLATE 54
Calla Leaves, about 1930
Zantedeschia aethiopica
See plate 8.

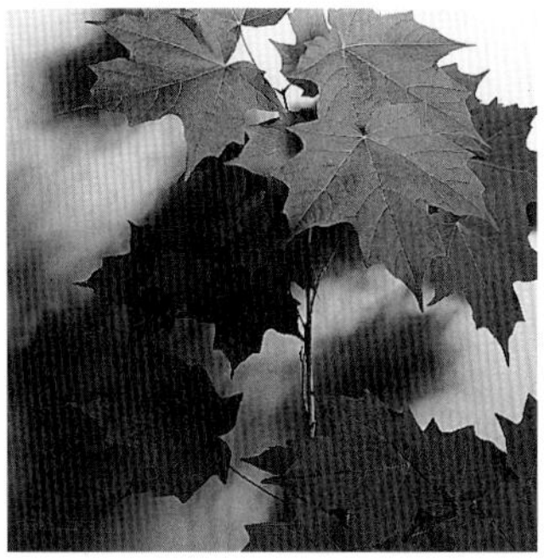
PLATE 55

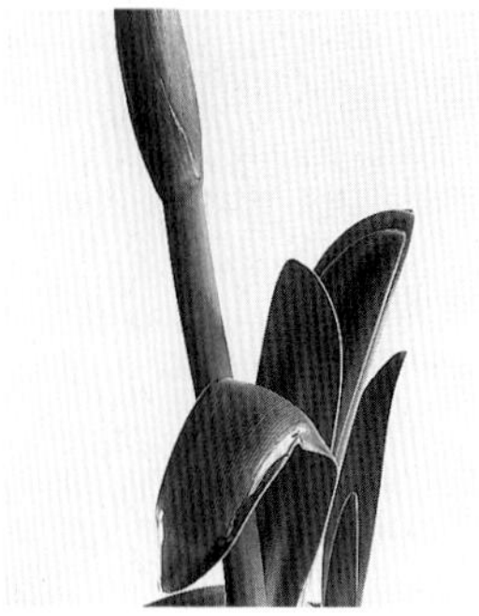
PLATE 56

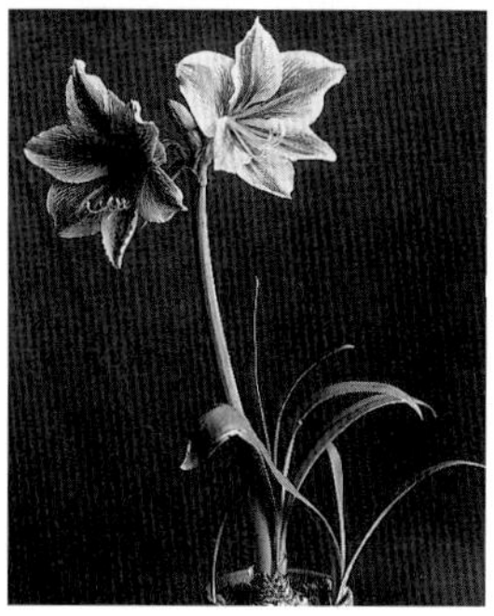
PLATE 57

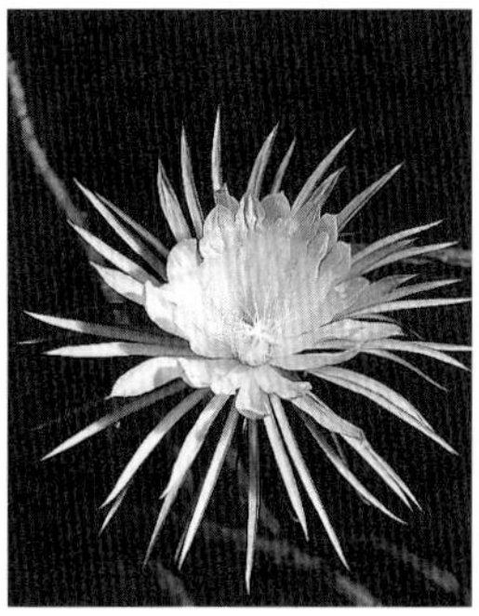
PLATE 58

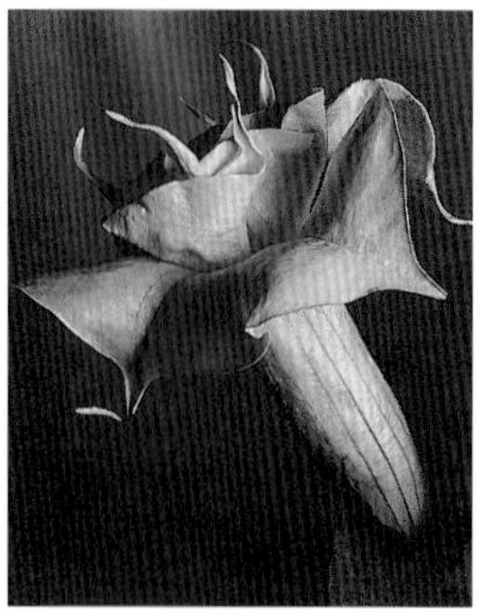
PLATE 59

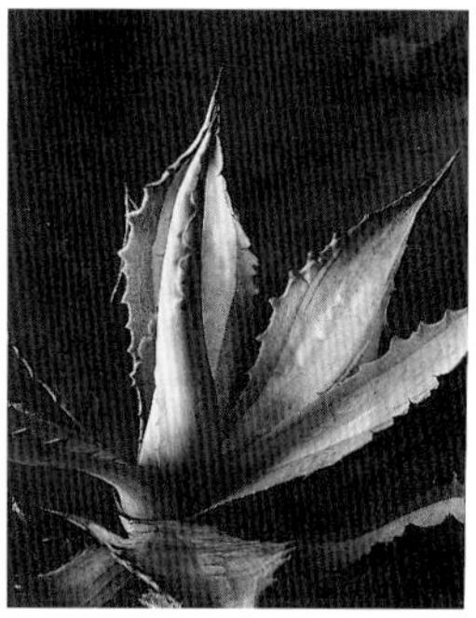
PLATE 60

PLATE 55

Maple Leaf Pattern for Screen, 1932

Acer macrophyllum

The Oregon maple, also called the bigleaf maple, is a stately tree that grows in western North America. It is well known in cultivation and has the largest leaves of all the maples. It grows to 80 feet tall or more. The 8- to 12-inch-wide leaves have five deep lobes. Yellow flowers on long pendulous racemes are produced in May. The bark of old trees is thick, rough, and deeply fissured. From Alaska to southern California, the species usually grows as individual trees in conifer forests, only occasionally forming forests of its own.

PLATE 56

Amaryllis, 1933

Hippeastrum hybridum exotica

A tropical native, the amaryllis is grown from a large bulb. From two to several flowers, up to 8 to 9 inches wide, form on the thick 2-foot-high stem. The leaves are broad and strap-shaped and sometimes appear only after the flower has blossomed. Many varieties are available in colors of white, pink, red, salmon, and orange. Some are marked or striped.

PLATE 57

Amaryllis Flower, 1933

Hippeastrum hybridum exotica

See plate 56.

PLATE 58

Flowering Cactus, about 1930

Selenicereus grandiflorus

Queen of the Night is a Mexican climbing cactus with thin gray-green branches. It clings by using its strong hooked tubercles. In summer, the large trumpet-shaped blooms open at dusk and last only until shortly after sunrise. The flowers are up to 12 inches in diameter and 8 inches tall, with white petals on the inside and golden petals on the outside. It exudes a rich vanilla fragrance.

PLATE 59

Datura, about 1930

Brugmansia candida

The datura is an evergreen shrub that produces tubular flowers and is often called angel's trumpet. It is a native of Peru and can grow to 10 to 15 feet tall. The leaves are dull green, 8 to 12 inches long. The flowers grow to 8 inches or more in length, in the shape of long, heavy white trumpets, single or double, and are very fragrant at night. Its flowers and seeds are poisonous.

PLATE 60

Agave, about 1930

Agave parryi variegata

The mescal agave forms stemless rosettes of stiff, wide, smooth gray leaves with straight end thorns. Clusters of creamy yellow flowers appear on a 12-foot-high spike when this plant is older.

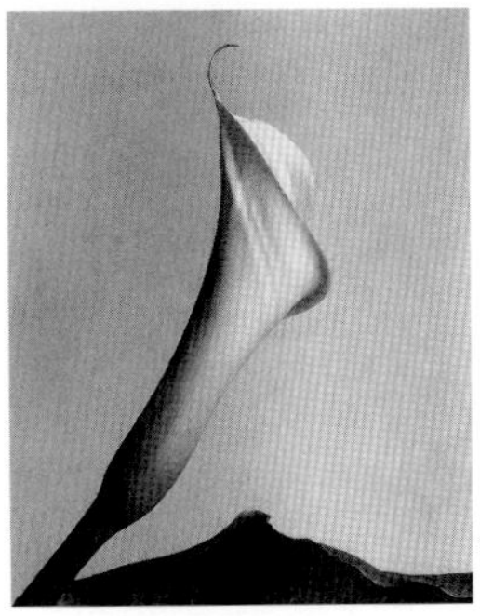

PLATE 61

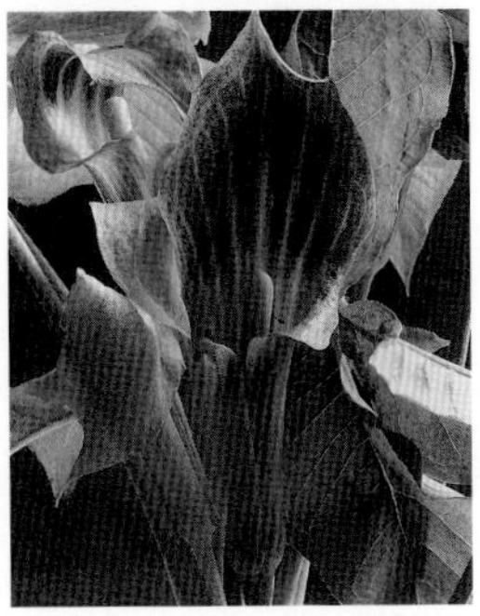

PLATE 62

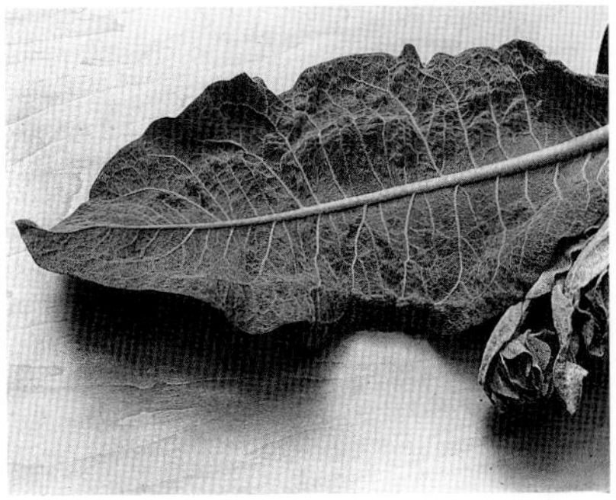

PLATE 63

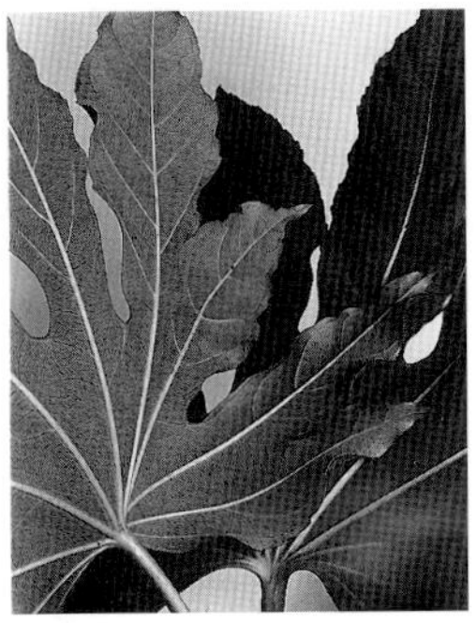

PLATE 64

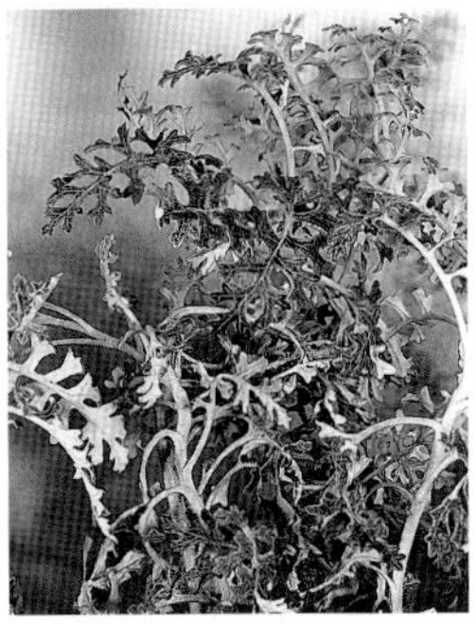

PLATE 65

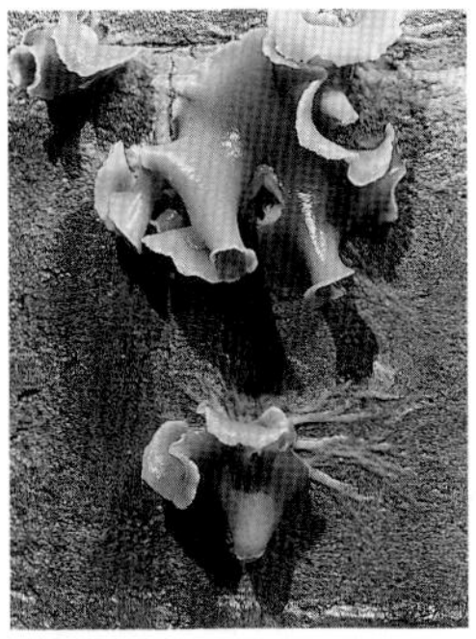

PLATE 66

PLATE 61
Calla with Leaf, about 1930
Zantedeschia aethiopica
See plate 8.

PLATE 62
Jack-in-the-Pulpit, Virginia, 1934
Arisaema triphyllum
Also called Indian turnip, the jack-in-the-pulpit is a perennial found in moist woods on the East Coast and in the South. The spathe is green with purple stripes and curves in a broad flap over the top of the spadix. The three-foliate erect leaves grow singly or in pairs.

PLATE 63
Tobacco, 1934
Nicotiana tabacum
Common tobacco is an annual narcotic herb or shrub that grows up to 5 feet high. It is closely furnished with huge, membranous, pale green leaves that are used in the manufacture of tobacco. The plant produces loose terminal clusters of fragrant, rosy, funnel-shaped flowers that open during daytime. Cultivated from early times by American Indians, tobacco often grows wild.

PLATE 64
Fatsia Japonica, 1930s
Fatsia japonica
The Japanese aralia is an evergreen shrub that grows 5 to 8 feet high. It has large, glossy, dark green leaves that are deeply lobed and fanlike and grow up to 16 inches wide on long stalks. During fall and winter, the plant produces clusters of small white flowers that become shiny black fruits.

PLATE 65
Dusty Miller, 1930s
Centaurea gymnocarpa
Dusty miller is a perennial that grows erect to 1 to 3 feet and has finely divided, white, softly textured leaves. It forms two or three heads of small purple flowers at the branch ends.

PLATE 66
Fungus in the Basement, Harbor View, 1930s
Peziza domiciliana
Found year-round throughout North America, this common fungus is usually associated with damp cellars, damp plaster, and rotten wood in houses. The fruit body is up to 4 inches across, at first gobletlike with a stem, then expanding and flattening with rolled edges. As it matures, its color changes from white to pale yellow-ochre to brownish buff.

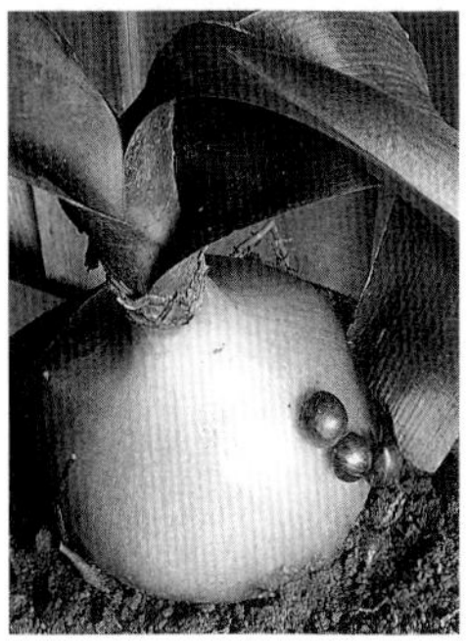

PLATE 67

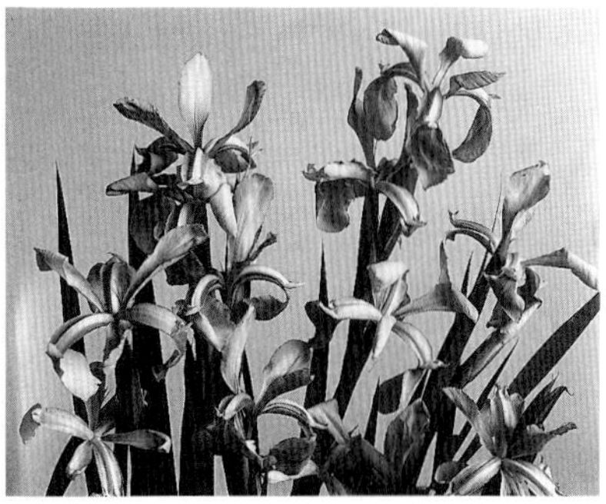

PLATE 68

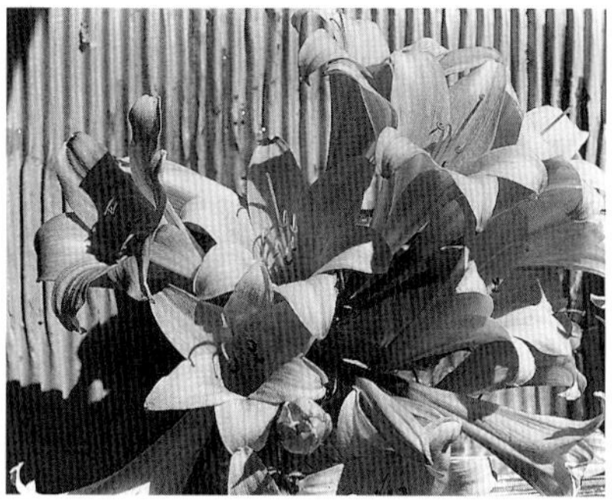

PLATE 69

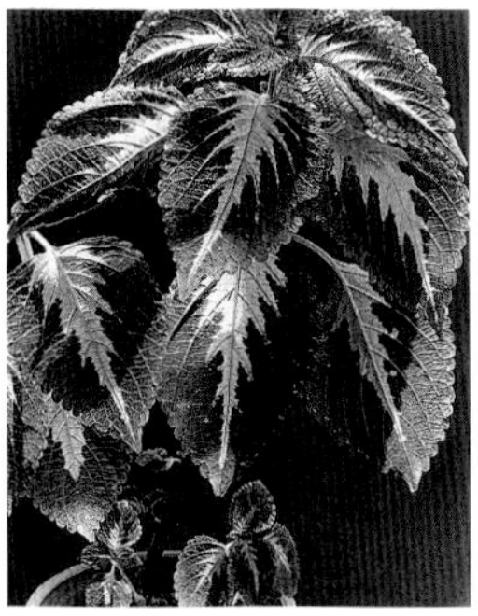

PLATE 70

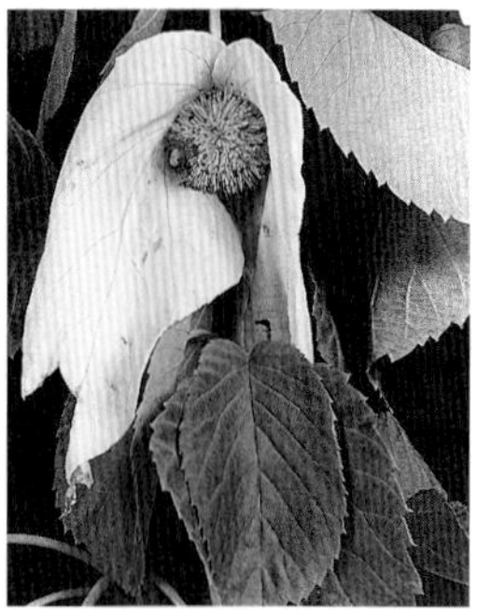

PLATE 71

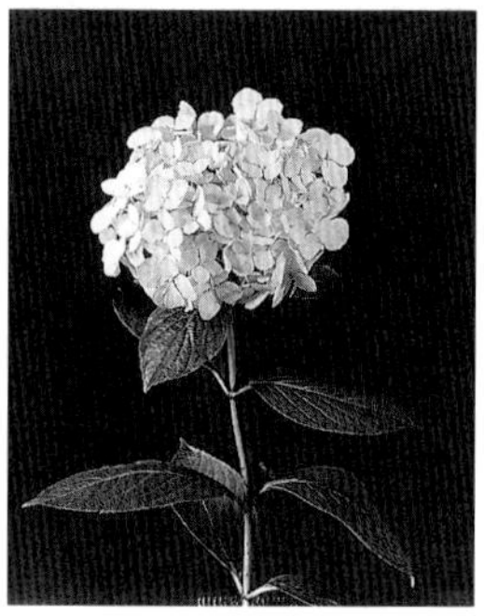

PLATE 72

PLATE 67
Pregnant Onion 2, 1934
Ornithogalum caudatum
The pregnant onion is a houseplant grown for its bulb and foliage, although it does produce tall stalks of green and white flowers. Its four to eight fleshy, strap-shaped leaves are 1½ to 2 feet long, ¾ to 1½ inches wide, and hang downward. The 3- to 4-inch-diameter gray-green bulb grows on top of the soil. It forms bublets under its skin which later drop out and root.

PLATE 68
Dutch Iris, 1930s
Iris versicolor
Dutch irises are grown from bulbs and have dainty, orchid-like flowers. Their stems are long and straight, 1½ to 2 feet tall, and their leaves are thin and narrow. The 3- to 4-inch-wide flowers have a broad color range including white, orange, purple, mauve, yellow-brown, and bicolors.

PLATE 69
Lilies, 1930s
Lilium regale
The regal lily is a Chinese species grown from a bulb and can reach 6 feet in height. The purple stem is covered with many short, narrow leaves and is topped by loose clusters of creamy white, funnel-shaped blooms with yellow centers and pale purple shading on the outside. The flowers, produced during the summer months, have a strong fragrance.

PLATE 70
Coleus, 1930s
Coleus hybridus
Coleus is an 18- to 24-inch-tall perennial usually grown as a houseplant. It has leaves 3 to 6 inches long that are brilliantly patterned and multicolored in shades of green, yellow, buff, orange, red, and purple.

PLATE 71
Davidia, 1930s
Davidia involucrata
The dove tree, a native of China, is a deciduous tree with a rounded crown that can grow to a height of 35 feet. The leaves are green, round to heart-shaped, and 3 to 6 inches long. When the tree flowers in late spring, it appears as if doves (or handkerchiefs) are resting on the leaves. The small, clustered, red-anthered flowers form between two differently sized, pendulous, white bracts, 4 to 6 inches long.

PLATE 72
Hydrangea, 1930s
Hydrangea macrophylla
The hydrangea is a deciduous shrub that can grow from 4 to 12 feet high. The leaves are thick, shiny, and toothed and grow to 8 inches long. The flowers form in large clusters ranging in colors of white, pink, red, or blue.

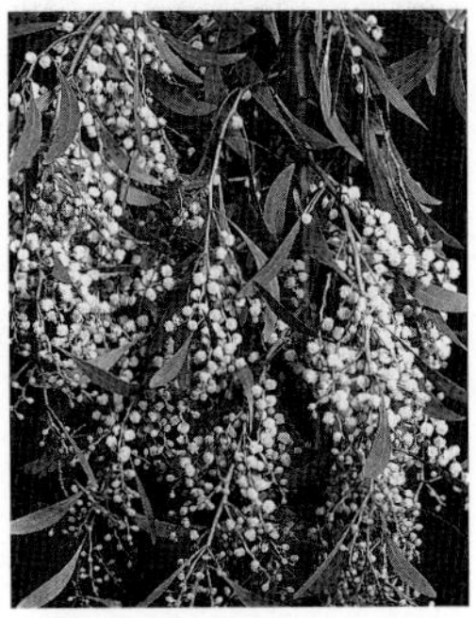

PLATE 73

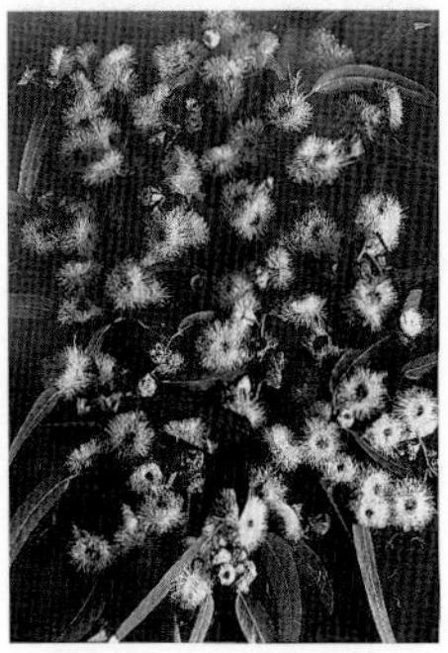

PLATE 74

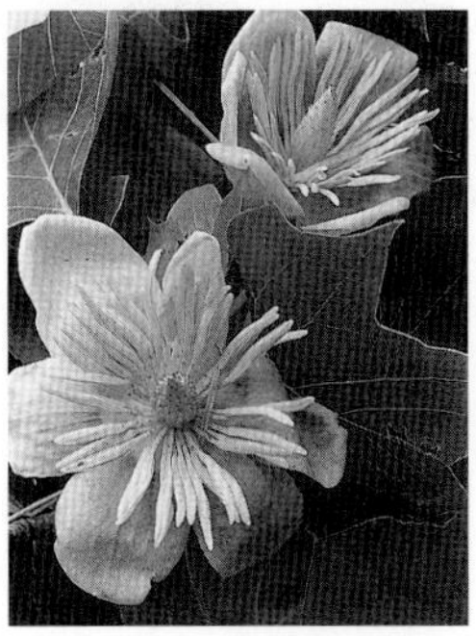

PLATE 75

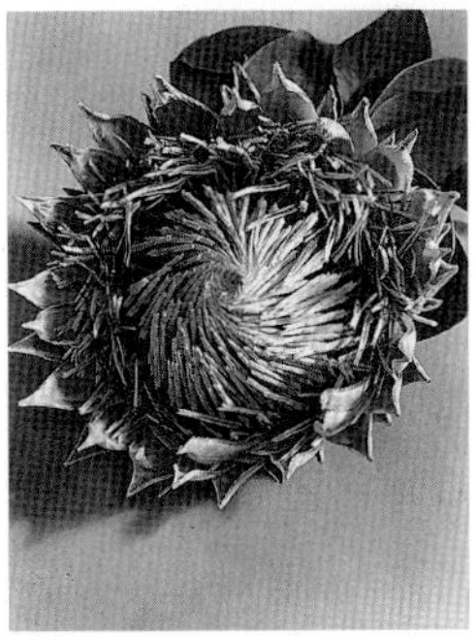

PLATE 76

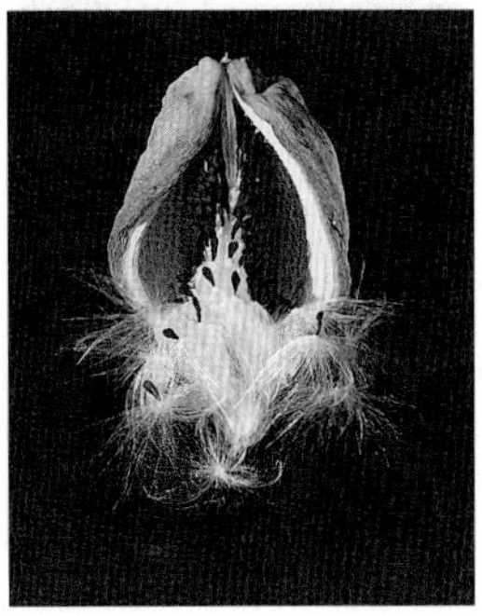

PLATE 77

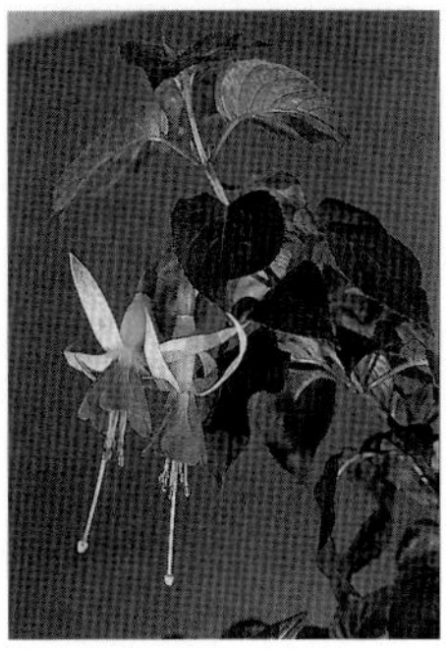

PLATE 78

PLATE 73

Acacia Flowers, 1930s

Acacia melanoxylon

The blackwood acacia is an evergreen tree that grows to 40 feet high and 20 feet in diameter. The leaves are dark green and 2 to 4 inches long. Its flowers are cream colored and form in short clusters during the spring.

PLATE 74

Eucalyptus Flowers, 1930s

Eucalyptus globulus

The blue gum eucalyptus is a grand, aromatic evergreen tree that can grow to 150 to 200 feet high. It has narrow sickle-shaped, dark green leaves, 6 to 10 inches long, and it characteristically sheds its bark. The flowers are creamy white and bloom in winter and spring.

PLATE 75

Tulip Tree 2, 1934

Liriodendrum tulipifera

The tulip tree is a deciduous tree native to the eastern United States. It grows to 60 to 80 feet high and up to 40 feet across to form a pyramidal shape. Its leaves are bright yellow-green, lyre-shaped, and 5 to 6 inches long and wide. In late spring, mature trees of at least ten to twelve years produce 2-inch-high tulip-shaped flowers that are greenish yellow outside and orange colored within.

PLATE 76

Blossom of Protea, 1935

Protea cynaroides

The king protea is a 3- to 5-foot-tall evergreen shrub native to South Africa. The leaves are oval and leathery. The flower heads are up to 12 inches across with white tubular flowers in a large cluster surrounded by pale pink to crimson bracts.

PLATE 77

Araujia Seed Pod, 1940

Araujia sericofera

The bladderflower, a native of Brazil, is a perennial vine in the milkweed family. It is a twiner that grows up into trees. The plant has slender stems that exude a milky sap. The arrow-shaped, dull green leaves are 2 to 4 inches long. The 1-inch-high flowers are white corollas topped by five lobes. The pear-shaped pod is 3 to 4½ inches long and splits when mature, revealing dark brown seeds topped with tufts of long silky hairs.

PLATE 78

Fuchsia, 1940

Fuchsia hybrida

The fuchsia is an evergreen shrub in frost-free climates and deciduous elsewhere. Its intensely colored flowers bloom from early summer until the first frost. The outer sepals are usually white, red, or pink, and the inner corolla, which can be single or double, is any of a range of colors including white, violet, pink, red, or orange.

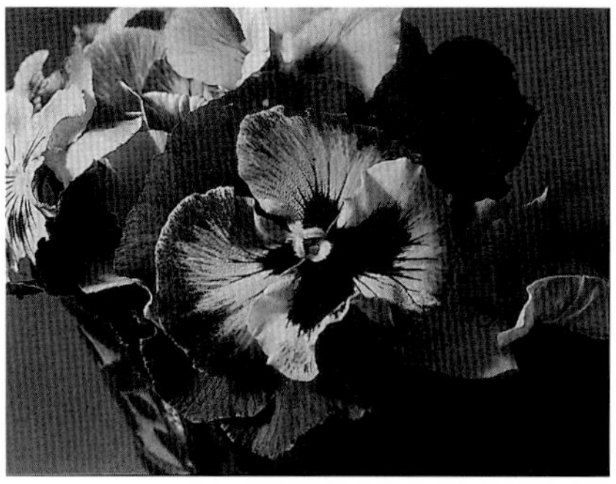
PLATE 79

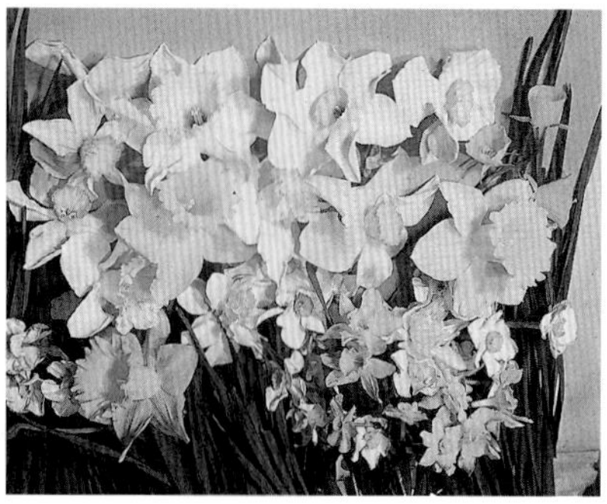
PLATE 80

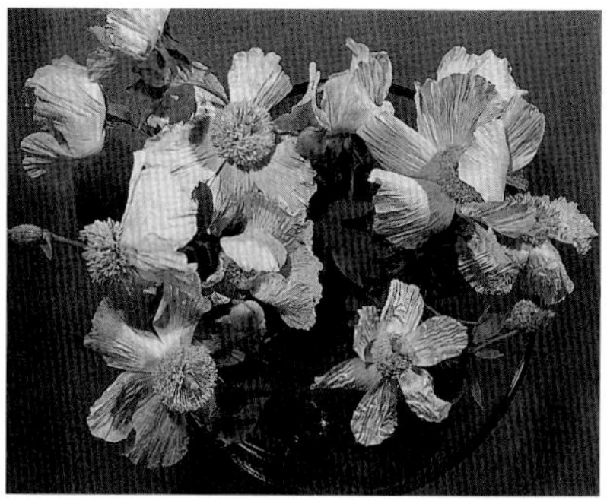
PLATE 81

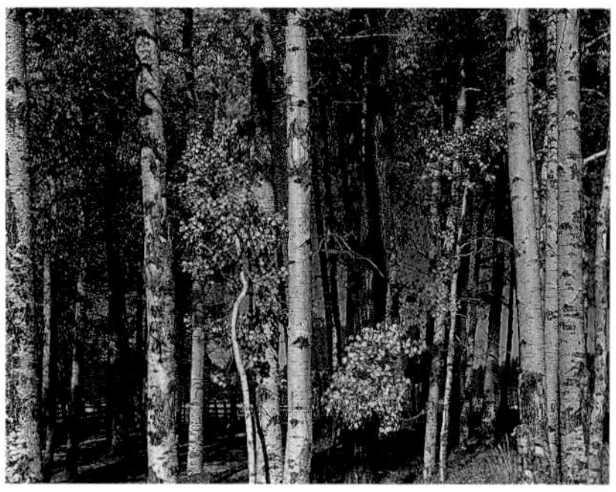
PLATE 82

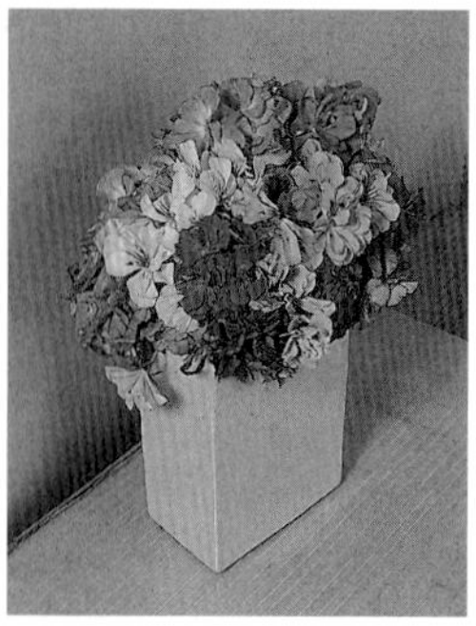
PLATE 83

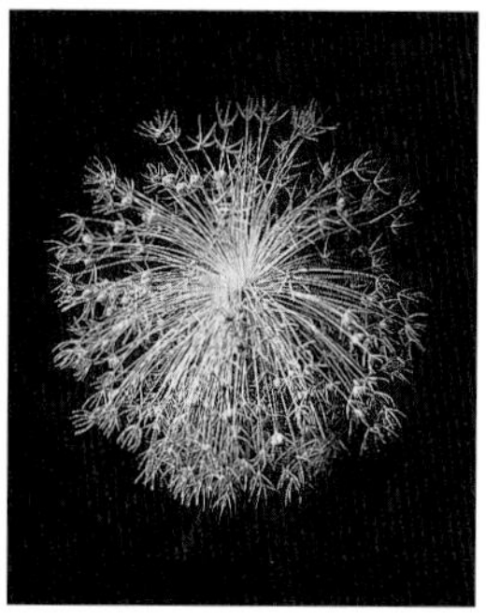
PLATE 84

PLATE 79
Pansies, 1940s
Viola wittrockiana
The pansy is an annual flower, 2 to 4 inches wide, in colors of white, blue, burgundy, rose, apricot, purple, or bicolors. The soft petals can be marked or patterned.

PLATE 80
Daffodils, 1940s
Narcissus (assorted varieties)
The spring-flowering daffodil is grown from bulbs. Its leaves are strap-shaped or rushlike. Flowers consist of rings of petals at right angles to the central trumpet or cup. Flowers are single or clustered and are usually shades of yellow and white, although other colors include orange, red, apricot, pink, and cream.

PLATE 81
Matilija Poppy, 1940s
Romneya coulteri
The matilija poppy is a perennial native to southern California and Mexico. It grows to 8 feet or more and has gray-green stems and deeply cut leaves. The fragrant flowers can be up to 9 inches wide with five or six white paperlike petals surrounding clusters of golden stamens.

PLATE 82
Aspens—Horizontal, about 1942
Populus tremuloides
The American aspen attains a maximum height of about 90 feet. The flattened leaf stalks allow the nearly circular leaf blades to quiver with the gentlest breeze. When young, its bark has a conspicuous whitish gray color. This aspen is believed to have the widest natural distribution of any North American tree.

PLATE 83
Geraniums in Variety, 1950s
Pelargonium hortorum
The common geranium is a shrubby plant that grows to a height of 3 feet or more. Its stem is succulent but becomes woody in older plants. The velvety leaves are round or kidney-shaped with scalloped edges. The clusters of 1- to 2-inch-wide flowers can be single or double and appear in solid colors of white, pink, rose, red, orange, and violet.

PLATE 84
Fireworks Plant, 1965
Allium giganteum
This ornamental allium grown from a bulb is a relative of the edible onion. It forms large, spherical clusters of lilac flowers on 5-foot-tall stems. The narrow leaves are 1½ feet long and 2 inches wide.

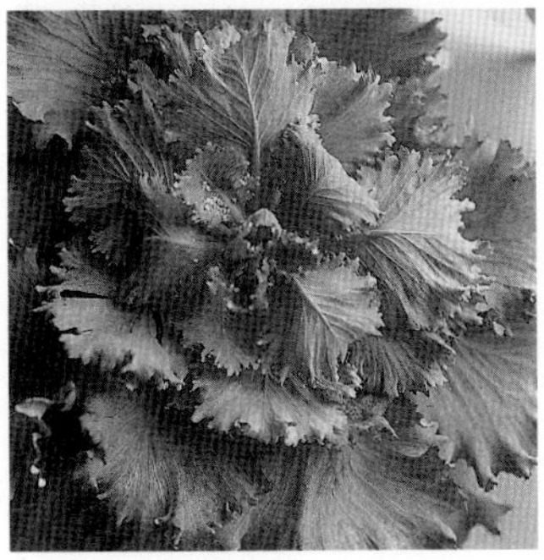

PLATE 85

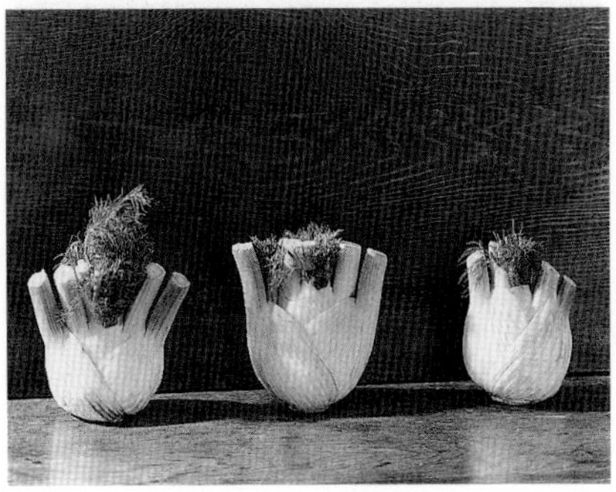

PLATE 86

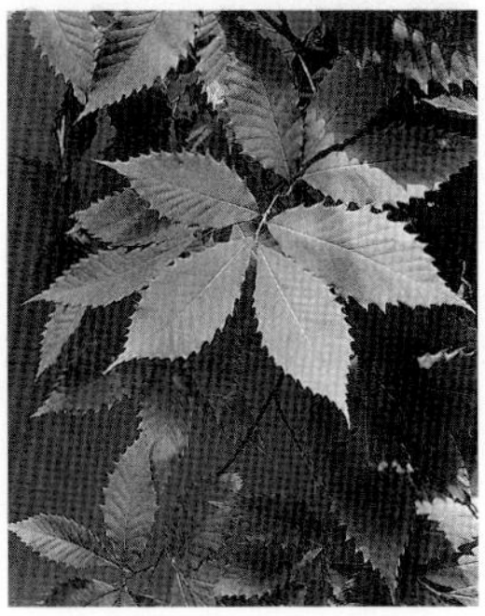

PLATE 87

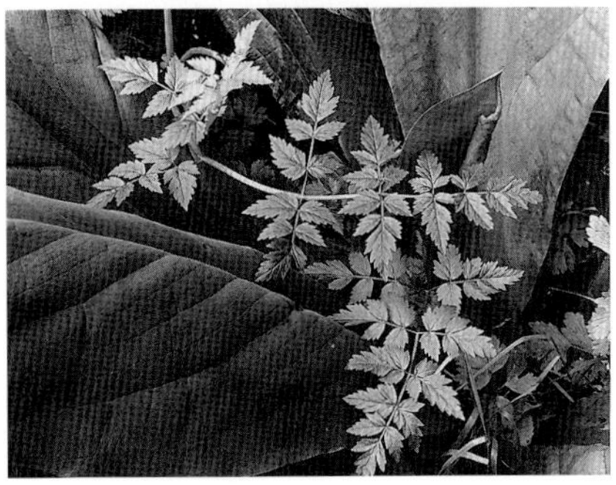

PLATE 88

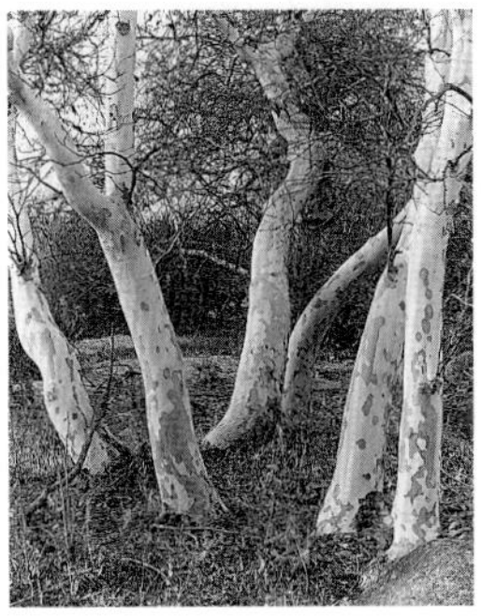

PLATE 89

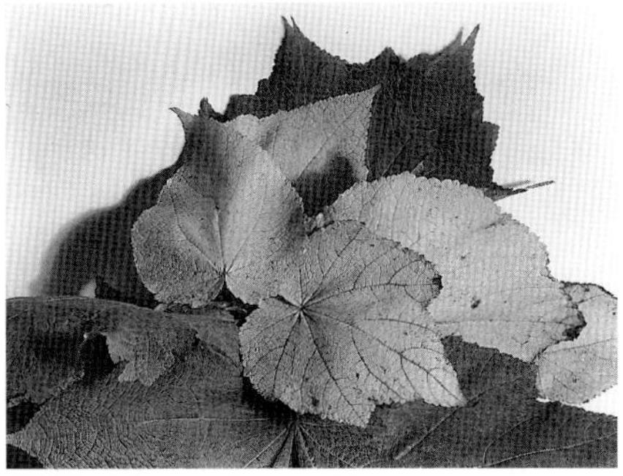

PLATE 90

PLATE 85

Ornamental Kale, 1976

Brassica oleracea

Ornamental kale has a flattish head of crisp and frilled leaves that are creamy white, pink, crimson, or purple. Often cultivated as a curiosity, it is a popular plant in exotic bedding schemes and as a table decoration. Although edible, it is not recommended for eating.

PLATE 86

Three Vegetables, 1946

Foeniculum vulgare

The common fennel, usually grown as a 3- to 5-foot-high summer annual, is used for its seeds. Its edible portion, the leaves and white fleshy bulbous stem, is eaten raw or cooked as a vegetable. A volatile oil, anethole, diffused throughout the plant and especially in the seeds, gives the plant its fragrance and flavor.

PLATE 87

Chestnut Leaves, 1940s

Castanea dentata

The American chestnut is now nearing extinction due to a fungus disease. Until 1930 it was the one American species of the four natives to grow into a large, shapely tree up to 100 feet tall. As a result of attacks of the chestnut blight, it rarely grows more than 40 feet high before succumbing, although some trees in western states have escaped the disease. The bark is smooth and dark gray and fissures into long, flat ridges, flaking at the edges. The leaves are 6 to 7 inches long with sharply toothed edges.

PLATE 88

Leaves, 1948

Oenanthe sarmentosa and *Lysichiton americana californicum*

An herbaceous wild parsnip with pinnate leaves, *Oenanthe sarmentosa* is a member of the carrot family and grows along streams or in marshes. *Lysichiton americana californicum,* commonly called skunk cabbage, is a swamp herb with stemless, large leaves forming from a thick, horizontal rootstock. It produces a boat-shaped spathe that is pale yellow and ill-scented.

PLATE 89

Sycamore Trees, 1949

Platanus racemosa

The American plane tree, or buttonwood, is often called sycamore. It has maplelike leaves and bark that flakes from the trunk in large, irregular patches. It is the most massive and probably the tallest deciduous tree in North America, reaching up to 170 feet high, with a trunk diameter of 14 feet and a massive irregular head of up to 110 feet in diameter.

PLATE 90

Sparmannia Africana, early 1950s

Sparmannia africana

The African linden is an evergreen shrub with dense, coarse foliage. In California, it is a fast grower that can reach 10 to 20 feet as a multitrunked thicket. The broad leaves are angled, up to 9 inches across, heavily veined, and have a velvety texture. From midwinter to early spring, it bears clusters of 1-inch white flowers with yellow stamens.

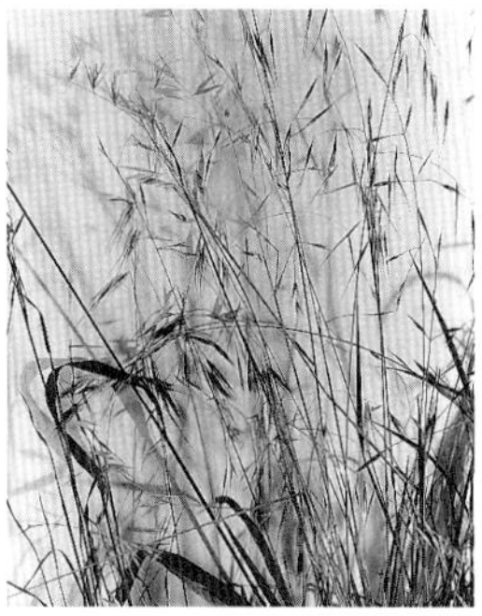
PLATE 91

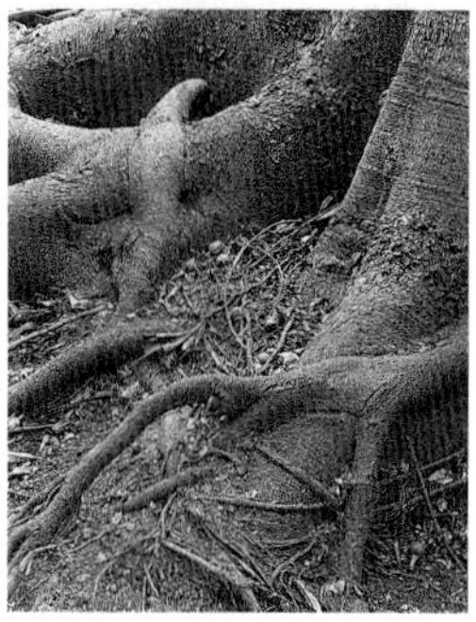
PLATE 92

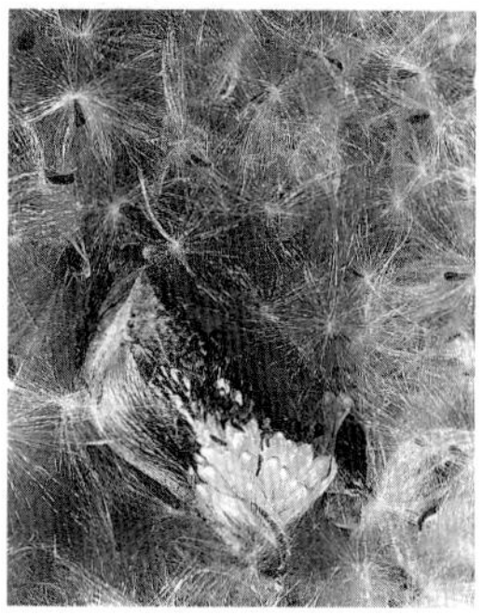
PLATE 93

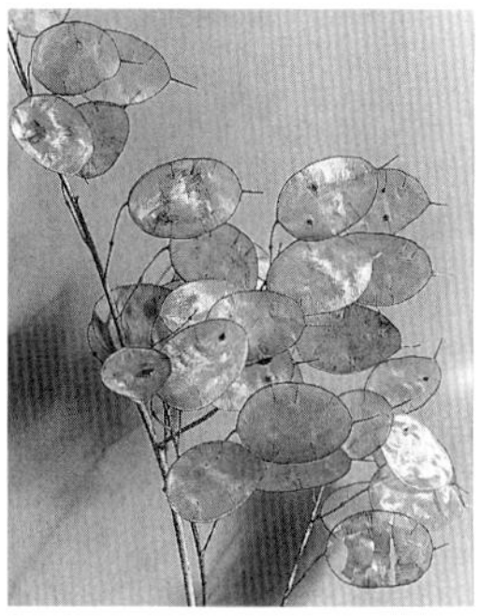
PLATE 94

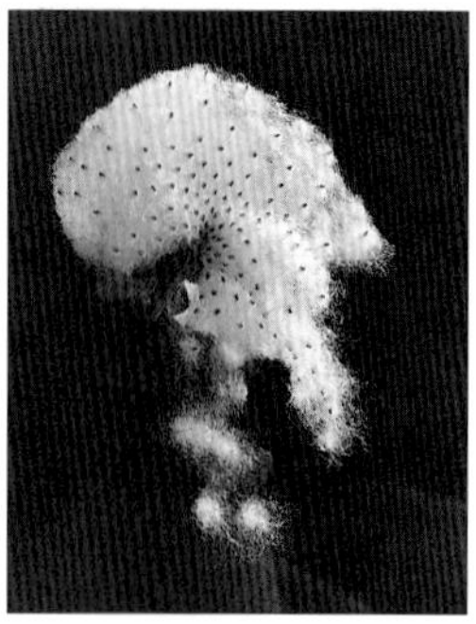
PLATE 95

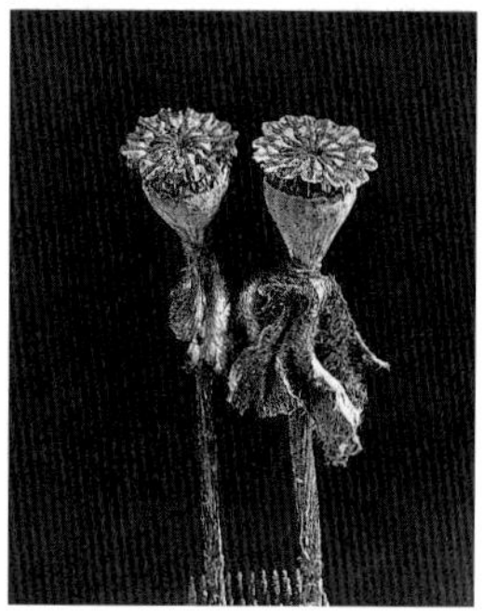
PLATE 96

PLATE 91
Grasses 2, 1952
Avena barbata
Slender wild oats is an annual grass that grows up to 3 feet high in spring and summer. It is a common weed in fields and waste places from Washington and Oregon to Arizona and California. The plant forms long-awned spikelets on weeping branches.

PLATE 92
Tree Roots, 1950s
Ficus macrophylla
The Moreton Bay fig is a huge evergreen tree that can reach up to 150 feet in height. It has oval-shaped, leathery leaves up to 10 inches long and 4 inches wide that are green above and brown below. Rosy leaf sheaths develop at the branch ends. The tree produces 1-inch-long, white-flecked, purple figs.

PLATE 93
Araujia, 1953
Araujia sericofera
See plate 77.

PLATE 94
Money Plant, 1956
Lunaria annua
The honesty plant, or money plant, is a biennial garden plant that grows up to 3 feet tall with coarse, heart-shaped leaves. It produces scentless violet flowers on long stalks that later form large flat seedpods. When the ripened pods drop their outer coverings and seeds, the remaining inner septums appear as silver circles, about 1¼ inches in diameter.

PLATE 95
Exploding Seed Pod, 1963
Anemone (unknown species)
Certain species of the anemone, also called the windflower, produce woolly seeds. The anemone is an attractive garden flower in a genus that contains about 85 species, all hardy perennials.

PLATE 96
Papaver Orientale, 1965
Papaver orientale
The Oriental poppy is a perennial that grows to 4 feet tall with divided leaves that are coarse and hairy. Its 3- to 6-inch-diameter flowers are single or double and range in color from pastel to brilliant shades. The ornamental mature seedpods are often used in dried flower arrangements.

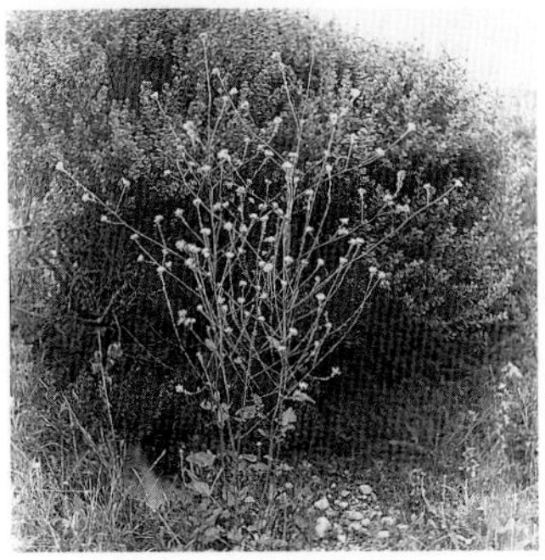

PLATE 97

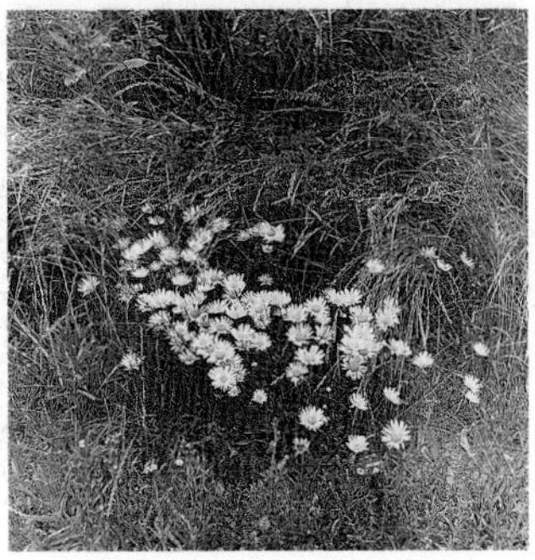

PLATE 98

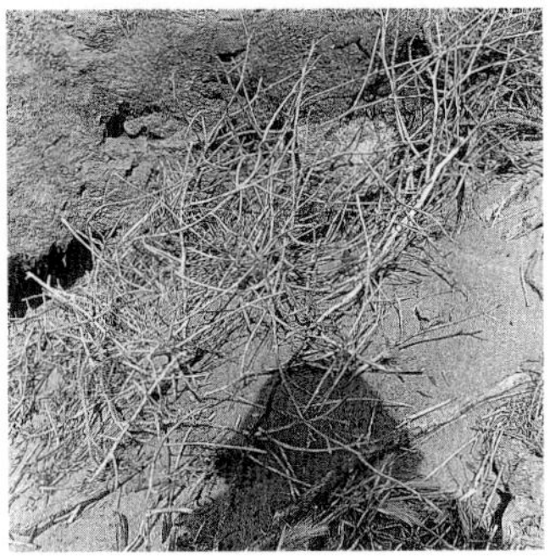

PLATE 99

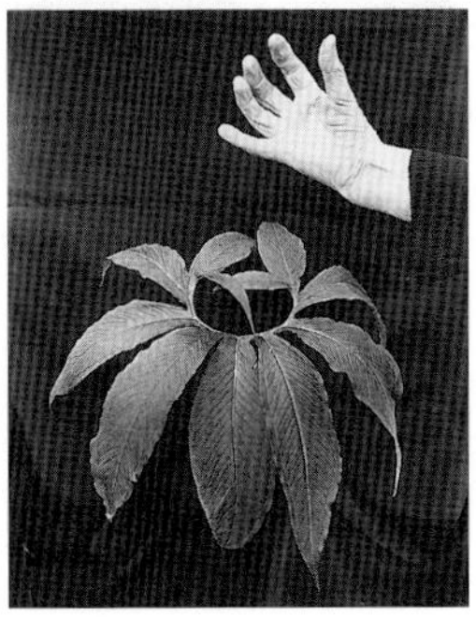

PLATE 100

PLATE 97

Wild Mustard, Pescadero Beach, 1966

Brassica arvensis

Wild mustard is an annual that grows to 1 to 2 feet, and occasionally much taller. The plant is dark green and rough with scattered stiff hairs. The lower leaves are deeply lobed or toothed, and the upper leaves are stemless and diamond-shaped. The bright yellow flowers are about ½ inch wide. Wild mustard, a European species, has become a common noxious weed in California fields and wastelands.

PLATE 98

Daisies, Pescadero Beach, 1966

Chrysanthemum maximum

The Shasta daisy is a hardy perennial that blooms during summer and fall. The plant is 2 to 4 feet tall and has coarse, lanceolate, leathery leaves. Its 2- to 4-inch-wide, marguerite-like flowers are white with gold centers.

PLATE 99

Self-portrait, Arcata, 1968

Cakile maritima

Sea rocket is an annual, fleshy, branching herb with purplish or white flowers. The genus has several species that are natives of sea and lake shores in North America and Europe. It grows prostrate or mound-forming to erect.

PLATE 100

Hand and Leaf of Voodoo Lily, 1972

Sauromatum guttatum

The voodoo lily is grown from a tuber. Its flower consists of a 12-inch-long greenish yellow bract accented with purple markings. A long, dark purple spike extends from the center. The bloom has a strong, repellent odor. The large, deeply lobed, fanlike leaf with 10-inch-long leaflets develops on a 3-foot-high stalk after the flower appears.

Chronology

1883 Born on April 12 in Portland, Oregon, to Isaac and Susan Cunningham.

ca. 1889 Moves to Seattle, Washington, with her family.

1903–7 Studies at the University of Washington, Seattle. Majors in chemistry.

1905-6 Acquires a 4-by-5-inch format camera from a mail-order correspondence school. Takes first photographs on University of Washington campus.

1907 Writes thesis, "The Scientific Development of Photography." After seeing the photographs of Gertrude Käsebier in the *Craftsman,* she decides to pursue photography as a career. Graduates from University of Washington.

1907–9 Works in Seattle portrait studio of Edward S. Curtis. Learns to retouch negatives and print with platinum paper.

1909 Travels to Dresden to study photographic chemistry at the Technische Hochschule. Visits the International Photographic Exposition in Dresden.

1910 Publishes her research on substituting lead salts for platinum in photographic printing paper in *Photographische Rundschau und Photographisches Centralblatt.* Visits Paris and London. In New York, visits Alfred Stieglitz's gallery "291." Returns to Seattle and opens portrait studio.

1913 Publishes "Photography as a Profession for Women" in Pi Beta Phi's journal, *The Arrow.*

1914 One-person exhibitions at the Brooklyn Institute of Arts and Sciences and at the Portland Art Museum, Oregon. Included in the International Exhibition of Pictorial Photography, New York. Illustrated review of her work published in *Wilson's Photographic Magazine.*

1915 Marries Roi Partridge on February 11. Son, Gryffyd, born December 18. Exhibits at Fine Arts Society, Seattle, with Roi Partridge, John Butler, and Clare Shepard. Included in Panama-Pacific International Exposition, San Francisco; Pittsburgh Salon of National Photographic Art; and Philadelphia Salon.

1917 Moves to San Francisco. Twin sons, Rondal and Padraic, born September 4.

1918 Works in San Francisco studio of Francis Bruguière. Meets Maynard Dixon and Dorothea Lange.

1920 Moves to Oakland, California, when Roi begins teaching at Mills College. Meets Edward Weston and Johan Hagemeyer.

1921 Resumes commercial portrait business. Creates first sharp-focus nature studies, at Point Lobos, and begins to photograph plant forms.

1922 Joins the Pictorial Photographers of America.

1923 Photographs light abstractions and begins series of magnolia studies.

ca. 1923 Makes first double-exposure photograph.

1928 Included in Pictorial Photographic Society exhibition, California Palace of the Legion of Honor, San Francisco.

1929 Exhibits at Berkeley Art Museum, California. Ten of her photographs included in *Film und Foto* exhibition, Stuttgart.

1931 One-person exhibition at the M. H. de Young Memorial Museum, San Francisco. Exhibits at Julien Levy Gallery, New York.

1932 One-person exhibition at the Los Angeles Museum. Exhibits with Group f.64 at the M. H. de Young Memorial Museum, San Francisco, where she was also included in *A Showing of Hands* exhibition.

1933 One-person exhibition at The Forum Hotel, Oakland.

1934 Included in *Leading American Photographers* exhibition, Mills College Art Gallery, Oakland. Marriage to Roi Partridge ends in divorce in June. Travels to New York City and Virginia. Photographs Alfred Stieglitz and begins documentary street photography.

1935 One-person exhibition at the Dallas Art Museum.

1936 One-person exhibition at the E. B. Crocker Art Gallery, Sacramento, California.

1937 Beaumont Newhall includes her in *Photography 1839–1937* exhibition at the Museum of Modern Art, New York. Photographs with 35mm camera.

1938 Begins photographing with 2¼-by-2¼-inch format cameras.

1940 Included in *A Pageant of Photography,* Golden Gate International Exposition, Treasure Island, San Francisco.

ca. 1940 Begins photographing in color for *Sunset* magazine.

1941 Included in *Photographers Exhibition* at Mills College Art Gallery, Oakland.

1943 Moves to Berkeley, California.

1947 Moves to San Francisco.

1947–50 Teaches at California School of Fine Arts, San Francisco.

1951 One-person exhibition at San Francisco Museum of Art.

1952 KRON-TV, San Francisco, produces documentary on Cunningham.

1953 One-person exhibition at Mills College Art Gallery, Oakland.

1954 Included in inaugural exhibition at Limelight Gallery, New York, and *Perceptions* at San Francisco Museum of Art.

1955 Included in Bay Area Photographers *San Francisco Weekend* exhibition at San Francisco Museum of Art. Interviewed by Herm Lenz, along with Dorothea Lange and Ansel Adams, in *U.S. Camera* article, "Interview with Three Greats."

1956 One-person exhibitions at Cincinnati Art Museum and at Limelight Gallery, New York.

1957 One-person exhibition at Oakland Art Museum.

1959 Included in *Photography at Mid-Century,* George Eastman House, Rochester, New York; one-person exhibition at Oakland Public Museum; and *The Photograph as Poetry,* Pasadena Art Museum.

1960 International Museum of Photography at George Eastman House, Rochester, New York, acquires large collection of her photographs. Travels to Berlin, Munich, Paris, and London.

1961 Travels to Norway, Finland, Sweden, Denmark, Poland, and Paris. Meets Man Ray. One-person exhibition at George Eastman House, Rochester, New York.

1964 One-person exhibitions at Chicago Art Institute and San Francisco Museum of Art. Becomes honorary member of American Society of Magazine Photographers. *Aperture* devotes its winter issue to her work. The Library of Congress acquires a large collection of her photographs.

1965 One-person exhibition at Henry Art Gallery, University of Washington, Seattle. Included in *Six Photographers 1965* at the College of Fine and Applied Arts, University of Illinois, Urbana.

1965–67 Teaches at San Francisco Art Institute.

1966 Featured in Fred Padula film, *Two Photographers: Imogen Cunningham and Wynn Bullock.*

1967 Elected a fellow of the American Academy of Arts and Sciences. One-person exhibition at Stanford Art Gallery, Stanford University, California.

1968 Receives honorary Doctor of Fine Arts degree from California College of Arts and Crafts, Oakland. One-person exhibitions at Museum of History and Technology, Smithsonian Institution, Washington, D.C.; Carl Siembab Gallery, Boston; and California College of Arts and Crafts, Oakland. Included in *North Beach and the Haight-Ashbury* exhibition, Focus Gallery, San Francisco. Teaches summer session at Humboldt State College, Arcata, California.

1970 One-person exhibitions at M. H. de Young Memorial Museum, San Francisco, and Witkin Gallery, New York. Receives a Guggenheim Fellowship to print her early negatives. John Korty films *Imogen Cunningham: Photographer.* University of Washington Press publishes *Imogen Cunningham: Photographs.* The Smithsonian Institution purchases a major collection of her work. San Francisco Mayor Joseph Alioto proclaims November 12 "Imogen Cunningham Day." *Album* magazine publishes a portfolio of her images.

1971 One-person exhibitions at Atholl McBean Gallery of the San Francisco Art Institute; 831 Gallery, Birmingham, Michigan; and Seattle Art Museum. *Creative Camera* publishes a portfolio of her images.

1972 Included in Group f.64 exhibition at University Art Museum, University of New Mexico, Albuquerque. One-person exhibition at Ohio Silver Gallery, Los Angeles.

1973 Teaches at San Francisco Art Institute. San Francisco Art Commission declares her "Artist of the Year." One-person exhibitions at The Metropolitan Museum of Art, New York; Witkin Gallery, New York; and San Francisco Art Commission "Capricorn Asunder" Gallery. Included in International Exhibition of Photography, Arles, France. *Images of Imogen* exhibition at Focus Gallery, San Francisco.

1974 Receives University of Washington's Alumnus Summa Laude Dignatus award. University of Washington Press publishes *Imogen!* Cunningham donates selected papers to the Archives of American Art, Smithsonian Institution. One-person exhibitions at Henry Art Gallery, University of Washington, Seattle, and The Oakland Museum.

1975 Creates The Imogen Cunningham Trust on February 14 to continue the preservation, exhibition, and promotion of her work. *Camera* magazine issues "Homage to Imogen" in October. One-person exhibitions at John Berggruen Gallery, San Francisco; Chevron Gallery, San Francisco; and Occidental Center Gallery, Los Angeles. Included in *Women of Photography* exhibition at San Francisco Museum of Art. Receives honorary Doctor of Fine Arts degree from Mills College, Oakland.

1976 Profiled in CBS documentary. One-person exhibitions at Stanford Art Gallery, Stanford University; Strybing Arboretum, San Francisco; and Grapestake Gallery, San Francisco. Dies on June 23.

Selected Bibliography

After Ninety. Introduction by Margaretta Mitchell. Seattle: University of Washington Press, 1977.

Berding, Christina. "Imogen Cunningham and the Straight Approach." *Modern Photography* 15, no. 5 (May 1951), pp. 36ff.

Cooper, Thomas Joshua, and Gerry Badger. "Imogen Cunningham: A Celebration." *British Journal of Photography Annual 1977* (1978), pp. 126–39.

Daniel, Edna Tartaul. "Imogen Cunningham: Portraits, Ideas, and Design" (transcript of 1959 interview). Berkeley, Calif.: Regional Cultural History Project, General Library, University of California, Berkeley, 1961.

Dater, Judy. *Imogen Cunningham: A Portrait.* Boston: New York Graphic Society, 1979.

The Eclectic Spirit: Imogen Cunningham. Exhibition catalogue. Essay by Richard Lorenz. Glasgow: Glasgow Museum, Kelvingrove, 1990.

Hall, Norman. "Imogen Cunningham: Sixty Years in Photography." *Photography* (London) 15 (May 1960), pp. 20–25.

Imogen Cunningham, Frontiers: Photographs 1906–1976. Exhibition catalogue. Text by Richard Lorenz. Berkeley, Calif.: The Imogen Cunningham Trust and United States Information Agency, 1987.

Imogen Cunningham: Photographs. Introduction by Margery Mann. Seattle: University of Washington Press, 1970.

Imogen Cunningham, Photographs 1921–1967. Exhibition catalogue. Foreword by Beaumont Newhall. Stanford, Calif.: Stanford Art Gallery, Stanford University, 1967.

Imogen Cunningham: The Poetry of Form. Exhibition catalogue. Introduction by Pradip Malde. Frankfurt: Fotografie Forum; Edition Stemmle, 1993.

"Imogen Cunningham: The Queen." *Newsweek,* October 21, 1974, p. 67.

Imogen! Imogen Cunningham Photographs 1910–1973. Introduction by Margery Mann. Seattle: Published for the Henry Art Gallery by the University of Washington Press, 1974.

"'Impressions in Silver' by Imogene [*sic*] Cunningham." *Los Angeles Museum Art News,* April 1932.

Internationale Ausstellung des Deutschen Werkbunds: Film und Foto. Exhibition catalogue. Stuttgart, 1929.

Jay, Bill. "Imogen Cunningham." *Album* (London) 5 (June 1970), pp. 22–38.

Kramer, Hilton. "Remembering Cunningham and White." *New York Times,* August 1, 1976.

Lenz, Herm. "Interview with Three Greats: Lange, Adams, Cunningham." *U.S. Camera* 18 (August 1955), pp. 84–87.

Lorenz, Richard. *Imogen Cunningham: Ideas without End, A Life in Photographs.* San Francisco: Chronicle Books, 1993.

———. *Imogen Cunningham: The Modernist Years.* Tokyo: Treville, 1993.

Maschmedt, Flora Huntley. "Imogen Cunningham—An Appreciation." *Wilson's Photographic Magazine* 51, no. 3 (March 1914), pp. 96ff.

Massar, Phyllis Dearborn. *Photographs by Imogen Cunningham.* Exhibition catalogue. New York: The Metropolitan Museum of Art, 1973.

Mozley, Anita Ventura. "Imogen Cunningham: Beginnings," in "Discovery and Recognition." *Untitled* 25 (1981), pp. 41–45.

[Osman, Colin]. "Photographs by Imogen Cunningham." *Creative Camera* (London) no. 85 (July 1971), pp. 222–31.

"Remarkable American Women 1776–1976." *Life,* special report, 1976, p. 43.

Rule, Amy, ed. *Imogen Cunningham, Selected Texts and Bibliography.* Essay by Richard Lorenz. Oxford: Clio Press, 1992.

White, Minor, ed. "Imogen Cunningham." *Aperture* 11, no. 4 (1964).

Acknowledgments

My continuing research on Imogen Cunningham and this book would not be possible without the trust, cooperation, and generosity of the entire Partridge family, especially Cunningham's three sons, Gryff Partridge, Padraic Partridge, and Rondal Partridge. As administrators of the Imogen Cunningham Trust, Rondal and Elizabeth Partridge, and their daughter Joan Partridge, have lovingly helped to prepare much of the material for *Imogen Cunningham: Flora,* and to them I am forever grateful.

Ed Marquand of Marquand Books has been an esteemed adviser and friend, and is responsible for bringing this book to publication. Suzanne Kotz has proved again to be an invaluable and exceptional fine-tuner of thoughts and words in editing this book. Rondal Partridge has printed with remarkable skill all the reproduction photographs from Cunningham's original negatives. M. Lee Fatherree has provided several superb copy photographs for text illustrations.

I would like to thank Robert G. Smith for his support, encouragement, and horticultural enthusiasm, as well as a few of my professors at Muhlenberg College who many years ago profoundly influenced my life: Dr. Francis Schaeffer, professor of botany, and Dr. Norma Iacovella and Dr. Ellen Callmann, professors of art history. A progression of people in my life have helped to develop my interests, attitudes, and experience, and to all of them I owe sincere thanks: the late Anna Lorenz Gigler, Rose Lorenz Hess, Caroline Keck and the late Sheldon Keck, Van Deren Coke, Nona R. Ghent, Robert McDonald, Judith Dunham, Edward De Celle, the late Billy Justema, and Evangeline J. Montgomery of the United States Information Agency–Arts America.

Professional thanks are due the many individuals who generously provided time and information to this project: Holly Forbes, Daniel Campbell, John Domzalski, Martin Grantham, Elaine Sedlack, and Kurt Zadnik, University of California Botanical Garden at Berkeley; Sarah Gates, University of Washington Herbarium; Drew Johnson and Janice Capecci, The Oakland Museum; Amy Rule and Dianne Nilsen, Center for Creative Photography, University of Arizona, Tucson; Scott Nichols; Anthony R. White; and Jean Wilton.

Information for the notes to the plates was gathered from the following sources:

Bailey, L. H. *The Standard Encyclopedia of Horticulture.* New York: Macmillan, 1925.

Bianchini, F. and F. Corbetta; Italia and Albert Mancinelli, trans. *The Complete Book of Fruits and Vegetables.* New York: Crown, 1976.

Dawson, E. Yale and Michael S. Foster. *Seashore Plants of California.* Berkeley: University of California Press, 1982.

Everett, Thomas H. *Living Trees of the World.* New York: Doubleday, 1977.

———. *The New York Botanical Garden Illustrated Encyclopedia of Horticulture.* New York and London: Garland Publishing, 1981.

Jepson, Willis Linn. *A Manual of the Flowering Plants of California.* Berkeley and Los Angeles: University of California Press, 1951.

Perry, Frances. *Flowers of the World.* London: Hamlyn, 1972.

Phillips, Roger. *Mushrooms of North America.* Boston: Little, Brown, 1991.

Robbins, W. W., Margaret K. Bellue, and Walter S. Ball. *Weeds of California.* Sacramento, Calif.: California State Department of Agriculture, 1970.

Van Gelderen, D. M. et al. *Maples of the World.* Portland: Timber Press, 1994.

Williamson, Joseph, ed. *Sunset New Western Garden Book.* Menlo Park, Calif.: Lane Publishing, 1979.